OUT OF SITE:
FICTIONAL ARCHITECTURAL SPACES

ORGANIZED BY ANNE ELLEGOOD

ESSAYS BY ANNE ELLEGOOD, RHONDA LANE HOWARD, AND MARK WIGLEY

NEW MUSEUM OF CONTEMPORARY ART
 NEW YORK   6.27.2002–10.13.2002
HENRY ART GALLERY, UNIVERSITY OF WASHINGTON
 SEATTLE   11.8.2002–2.2.2003

D1824511

PUBLISHED BY THE NEW MUSEUM OF CONTEMPORARY ART IN COLLABORATION WITH THE HENRY ART GALLERY, UNIVERSITY OF WASHINGTON, SEATTLE.
COPYRIGHT © 2002, NEW MUSEUM OF CONTEMPORARY ART. ALL RIGHTS RESERVED.
DESIGNED BY PURTILL FAMILY BUSINESS
PRINTED BY CRW GRAPHICS, NEW JERSEY
DISTRIBUTED BY D.A.P.

ISBN 0-915557-85-1

THIS PUBLICATION IS MADE POSSIBLE BY THE PENNY MCCALL PUBLICATIONS FUND AT THE NEW MUSEUM. DONORS TO THE PENNY MCCALL PUBLICATIONS FUND ARE JAMES C.A. AND STEPHANIA MCCLENNEN, JENNIFER MCSWEENEY, ARTHUR AND CAROL GOLDBERG, DOROTHY O. MILLS, AND THE MILLS FAMILY FUND.

ALTOIDS IS THE PRESENTING SPONSOR OF OUT OF SITE AT THE NEW MUSEUM OF CONTEMPORARY ART.

ADDITIONAL SUPPORT IS PROVIDED BY THE TOBY DEVAN LEWIS FUND FOR EMERGING ARTISTS, AND SCOTT J. LORINSKY.

COVER IMAGE: HALUK AKAKÇE
STILL LIFE, 2002
INSTALLATION WITH TWO-CHANNEL VIDEO PROJECTION
DIMENSIONS VARIABLE
COURTESY OF BERNIER/ELIADES, ATHENS, AND DEITCH PROJECTS, NEW YORK

NEW MUSEUM

1

ANNE ELLEGOOD is Associate Curator at the New Museum of Contemporary Art where she has organized *Candice Breitz: Babel Series*; *David Galbraith & Teresa Seeman: Waveform*; *Kristin Lucas & Joe McKay: The Electric Donut*; and *Cin-o-matic: Memory and Cinematic Experience*, a group show examining the impact of new technologies on the cinematic experience, co-organized with Michele Thursz. Among other projects are *Transparent Architecture*, exploring surveillance and its relationship to architectural spaces, and *The Meaning of Style*, a show about the lifestyle and design associated with particular subcultures. She received her MA in Curatorial Studies from the Center for Curatorial Studies at Bard College in 1998.

RHONDA LANE HOWARD is Associate Curator at the Henry Art Gallery where she has curated the exhibitions *Cheap Thrills: Great Design for $5 or Less*; *Jeffry Mitchell: Hanabuki*; *Ernesto Neto's Flying Gloup Nave*; *Prints by Painters: American Abstractions*; *Shifting Ground: Transformed Views of the American Landscape*; and *Inigo Manglano-Ovalle: Banks in Pink and Blue*. She has written exhibition publications and presented lectures since 1996. Prior to her tenture at the Henry she was research associate at the Corcoran Gallery of Art, Washington D.C., from 1994 to 1996, and a Curatorial Assistant at the Fisher Gallery at the University of Southern California, Los Angeles, from 1992 to 1994.

MARK WIGLEY is Professor of Architecture at Columbia University. He is the author of *The Architecture of Deconstruction: Derrida's Haunt* (1993), *White Walls, Designer Dresses: The Fashioning of Modern Architecture* (1995), and *Constant's New Babylon: The Hyper-Architectures of Desire* (1988). He co-edited *The Activist Drawing: Situationist Architectures from New Babylon to Beyond* (2001), and is currently working on a prehistory of virtual space.

FOREWORD

Out of Site is exactly the kind of exhibition the New Museum should be doing. It is a thematic presentation of emerging artists focused on a subject that is timely and urgent. Throughout its twenty-five year history, the Museum has produced thematic exhibitions that have distinguished the Museum's exhibition program around the world. *Out of Site* continues that tradition. Because of our size, we can be flexible and responsive to new developments in the field, and *Out of Site* advances our mission to seek out the new and adventurous.

    Out of Site has been organized by Anne Ellegood, Associate Curator. She proposed the idea after observing how artists today are grappling with changing notions of space in our technological world. *Out of Site* brings together fifteen artists (including one collaborative) who create fictional architectural spaces in active dialogue with a world transformed by digital, cyber, and virtual experiences. Ms. Ellegood has organized several important exhibitions for the New Museum's Media Lounge, but this is her first full-floor effort; she brings to it specific knowledge gained through her previous exhibition work focusing on new technology as well as her keen interest in young artists at work today. The work in the exhibition has tremendous texture and drama, and demonstrates the far-reaching implications of (digital) technology on artists of all persuasions—not just those working in "new media."

    This book brings together several useful and complementary viewpoints for considering the artists included. Architectural historian Mark Wigley has contributed an outstanding text on the speculative science fiction visions of contemporary architects and their historical precedents. Rhonda Lane Howard from the Henry Art Gallery has provided a discussion on the obsolete dialectic between the real and the virtual and the many philosophical perspectives it has engendered. Finally, Anne Ellegood has given us a superb and detailed description of the concerns of the participating artists as expressed in the work on view.

    I am grateful to our colleagues at the Henry Art Gallery—Richard Andrews, Director; Elizabeth Brown, Chief Curator; and Rhonda Lane Howard, Associate Curator—for their

enthusiastic response to the subject, for contributing to the catalogue, and for hosting the exhibition in Seattle.

    Finally, the artists are helping to reveal through their own special and private processes, how drastically transformed our landscape has become and how permeable the real and virtual are in the twenty-first century. One of the earliest exhibitions I organized in the mid-1970s was called *Architectural Analogues*, and though it sounds like it could allude to the digital, it was about fictional, fantastic architecture created through sculptural models and shifts in scale that had little, if anything, to do with new technologies. In retrospect, the concept seems circumscribed by comparison to the far-ranging discussions about spatial configurations and metaphors that have emerged in the intervening quarter-century.

    As Ms. Ellegood points out, we are at the threshold of a new world—initiated by the third (digital) industrial revolution—which raises fundamental issues about the nature of reality and our representation of it.

— LISA PHILLIPS, HENRY LUCE III DIRECTOR

ACKNOWLEDGMENTS

Out of Site was inspired by the outstanding work of the artists included. I would like to begin by thanking each of them for their genuine insight into the impact of particular cultural phenomena on our understandings of space and the incredible range of visual languages and formal articulations their ideas have taken through the depiction of architectural sites. It has been a great honor to work with each of them and present their work. I am grateful to the following dealers and collectors who have put their time and energy into facilitating loans or generously lending their works to the exhibition: Mary Jane Aladren of Nylon Gallery, Adam Ames, Shoshana Blank of Shoshana Wayne Gallery, Mary Boone & Ron Warren of Mary Boone Gallery, Spencer Brownstone, Elizabeth Burke & Abby Messitte of Clementine Gallery, Douglas S. Cramer, Dimitri Daskolopoulos, Edith and Joseph De Chiara, Laurie De Chiara, Jeffrey Deitch, Derek Eller, Susan Evans, Hugh J. Freund, Janice Guy & Margaret Murray of Murray Guy Gallery, Peter Halley, Christian Haye & Jenny Liu of The Project, Paul Kasmin, Andrew Kreps, Sara Meltzer & Adam Frank of Sara Meltzer Gallery, Paul Pankauski, Andrea Rosen & John Connelly of Andrea Rosen Gallery, and Henry Urbach & Jonathan Allen of Henry Urbach Architecture. For their generous financial support, I want to thank the Penny McCall Publications Fund, including donors James C.A. and Stephania McClennan, Jennifer McSweeney, Arthur and Carol Goldberg, Dorothy O. Mills, and the Mills Family Fund; the Toby Devan Lewis Fund for Emerging Artists; Scott J. Lorinsky; and Altoids, the presenting sponsor, all of whom have a sustained and admirable commitment to emerging artists.

It has been a great pleasure working with Rhonda Lane Howard, Associate Curator at the Henry Art Gallery, who immediately responded enthusiastically to the concept of the show, worked very hard to make sure that it would travel to the Henry, and wrote a perceptive and enjoyable text for the catalogue in record time. The Henry Art Gallery is an ideal institution to present *Out of Site*, and I want to extend my thanks to Richard Andrews and Elizabeth Brown at the Henry for their support. I am thrilled to have Mark Wigley's fantastic contribution to the catalogue; I will be forever grateful that he finally relented after I kept insisting that he indeed *did* have time to write a text. I am also indebted to Tim Yohn for his careful edits. Conny Purtill's exceptional catalogue design is indicative of his adeptness at creating both an elegant and meaningful visual counterpoint to the exhibition and a gorgeous book that can stand alone. Many thanks must also go to his partner Jenelle Porter who kept us on track with a perfect balance of support, humor, and firmness.

We have an extraordinary staff at the New Museum, and *Out of Site* could not have been realized without the hard work of numerous individuals. Although it isn't possible to mention everyone by name, my appreciation goes out to the entire staff. I feel privileged to work with Dan Cameron whose energy, enthusiasm, and unwavering love of art is a constant inspiration to me. I want to thank Lisa Phillips for all her guidance and support; her generosity of spirit and vast experience has been invaluable. Simply observing Dan and Lisa at work has given me some of my most valuable lessons in curatorial practice. Both Dennis Szakacs and Johanna Burton provided insightful and articulate suggestions for my text, while simultaneously playing a large part in keeping my sense of humor intact. Melanie Franklin's keen eye also contributed greatly to the catalogue. Jennifer Ray's ability to juggle countless details while remaining perfectly calm has been an incredible comfort; Peter Gould's accomplished understanding of what it takes to get a show installed has made him a huge asset; and we all know that nothing would be accomplished at the Museum without John Hatfield. There have been several outstanding interns in the curatorial department who have contributed greatly throughout the organization of the show, and I'd like to specifically acknowledge Mary Ellington, Luiza Interlenghi, and Tim Popa. Lastly, I want to thank Marcia Tucker, founding director of the New Museum, with whom I had the profound pleasure of working directly out of graduate school; she continues to be both a generous and enlightened mentor as well as a great friend.

— ANNE ELLEGOOD, ASSOCIATE CURATOR

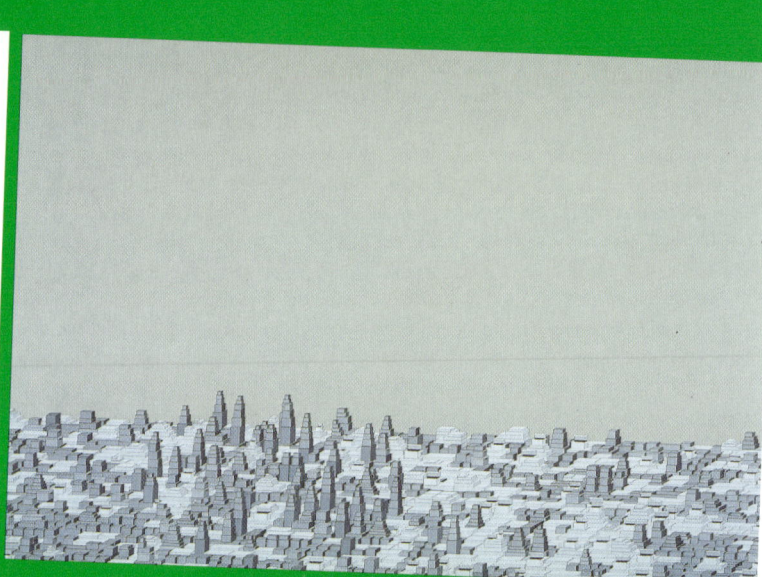

MATTHEW NORTHRIDGE
UNTITLED, 2002
COLLAGE
30 X 44 INCHES UNFRAMED
COURTESY OF THE ARTIST
PHOTO: JASON MANDELLA

In his 1984 science fiction novel *Neuromancer*, William Gibson repeatedly describes the highly advanced technologies of the societies in which action takes place in architectural terms. In these future societies, technologies "penetrate the bright walls of corporate systems, opening windows into rich fields of data."[1] An architecture of matrixes and grids forms the foundation of Gibson's world, supporting lightning-speed communication and extensive transportation channels and allowing for the continual negotiation between the physical and virtual in the face of personal risk and the ever-present threat of a chaotic breakdown.

Gibson coined the term "cyberspace" for the non-physical world his protagonists inhabited and the term resonated so profoundly with techies, theorists, and pop culture enthusiasts that it quickly entered the mainstream. His description of cyberspace concludes with a metaphor of the city:

A consensual hallucination experienced daily by billions of legitimate operators, in every nation.... A graphic representation of data abstracted from the banks of every computer in the human system. Unthinkable complexity. Lines of light ranged in the nonspace of the mind, clusters and constellations of data. Like city lights, receding....[2]

Gibson's cyberspace takes place in the mind, "a consensual hallucination," reminding us that the creation of fantastical architectural sites and cityscapes has long held the imaginations of writers, artists, and architects, from Thomas More's 1516 novel *Utopia* to architect Bruno Taut's expressionistic proposals early in the twentieth century, to contemporary art works such as Chris Burden's sprawling, Lilliputian sculptural topographies and Thomas Demand's photographic practice rooted in the architectural model. Efforts to understand space and perspective are commonly played out through architecture, a discipline inherently linked to spatial considerations. Further, architecture has historically been understood as a physical manifestation of psychological states of mind, needs, and desires often brought on by social changes and innovations in industrialization and technology. As exaggerated and complicated as they are, Gibson's sci-fi environments come straight out of lived experience.

The technological advances of recent years and ever-increasing access to digital innovation have spawned a profound cultural shift, permanently altering the way we experience and

JULIE MEHRETU
RETOPISTICS: A RENEGADE EXCAVATION, 2001
INK AND ACRYLIC ON CANVAS
96 X 216 INCHES
DIMITRI DASKOLOPOULOS COLLECTION, ATHENS
COURTESY OF THE PROJECT, NEW YORK AND LOS ANGELES

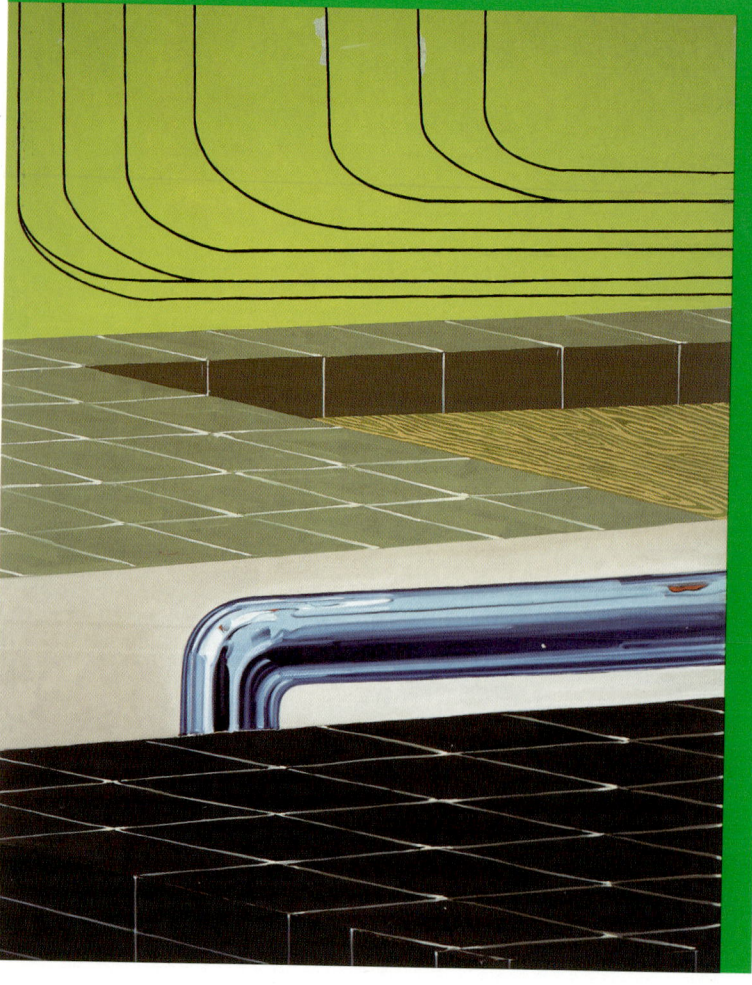

CANNON HUDSON
SERPENT, 2002
OIL ON CANVAS
72 X 54 INCHES
COURTESY OF THE ARTIST
PHOTO: JASON MANDELLA

represent space. Evidence of this shift is abundant within the electronic experiences now integrated into daily life, which are marked by speed of mobility through space, the viewing of multiple perspectives simultaneously, overlapping windows of data and visual information, non-linear hypertexts, the breakdown of physical boundaries and temporalities, morphing forms, and navigation through interstitial spaces between seemingly infinite visual representations. Typically considered opposites, today virtual and real space influence each other to such a degree that a dialectical discourse no longer addresses the phenomenon adequately.

The artists in *Out of Site* create fictional architectural spaces and topographies. What gives these diverse works a special relevance is that they use architectural constructs to investigate how the growth of digital culture, virtual reality, and global expansion have impacted our culture. Population growth, migration, and sub/urban sprawl taking place on a global platform have also been catalysts to a reevaluation of spatial relationships, just as new patterns of colonization and trade, powered by a global economic system, have affected architectural forms and changed our relationships to the spaces we inhabit. At the heart of the exhibition is a shift in the understanding of perspective and spatial relationships by visual artists as significant within the history of art as Filippo Brunelleschi's discovery of one-point perspective during the early Renaissance, enabling artists for the first time to accurately depict real space in a two-dimensional format, and the cubists' invention of a pictorial space that presented multiple perspectives of a single object simultaneously. As Elizabeth Grosz surmises, "Perhaps the most striking transformation effected by these technologies is the change in our perceptions of materiality, space, and information, which is bound directly or indirectly to affect how we understand architecture, habitation, and the built environment."[3] Not surprisingly, contemporary artists have turned to architecture to investigate the world in which we now live and to make projections about the future.

The spaces depicted in the exhibition are fictional in that they do not refer to existing sites, yet they are manifestations of today's experiences of navigating space—both physically, in our homes and workplaces, as well as in the digital realms of the Internet, video games, and other virtual environments—and are informed by an awareness of the changes in architectural discourse and the design of buildings, as well as the integration of technologies into sites themselves in recent years. The use of the term "fictional" here is self-consciously aligned with the claim of several theoreticians, including Jean Baudrillard, Paul Virilio, Celeste Olalquiaga, Pierre Levy, and Edward W. Soja among others, that we can no longer distinguish the "real" from the "virtual," or simply don't find the distinctions meaningful anymore. The environments in *Out of Site* share characteristics with Iain Chambers's description of the city when he wrote, "The city exists as a series of doubles; it has official and hidden cultures, it is a real place and a site of imagination.... We discover that urban 'reality' is not single but multiple, that inside the city there is always another city."[4] Computer design tools and the Internet offer an infinitely diverse space, directly impacting urban configuration and organization. Architecture is now characterized by its adaptability and flexibility. Temporary sites are accepted as valid; the impulse toward monumentality is receding. Digital culture and global economic markets participate in the deterritorialization of space. Boundaries between interior and exterior, local and global have blurred and given way to hybrids of all kinds. Liminal spaces existing between physical structures and perceived planar fields of distinct dimensions are consistently sites of investigation. Architectural theorist Anthony Vidler describes today's experience as that of "occupying" space, reflected in an architecture with shifting scales, transmutability, and a process of dismantling.[5] The artists in *Out of Site* respond to all these changes by organizing and articulating space and creating fantastic new forms as multifarious as the culture that has spawned them.

Less than twenty years after his novel was written, aspects of Gibson's world of electronic interconnectedness, virtual reality travel, and the struggle for physical space in claustrophobically overpopulated metropolises, no longer sound otherworldly. This fact highlights science fiction's role as "an elaborate shock-absorbing mechanism," to use the words of Fredric Jameson, preparing us for the changes that will one day infiltrate our lives.[6] But more important from an ideological perspective is what science fiction, particularly the subgenre of the utopic or dystopic sci-fi novel, tells us about ourselves. Multimedia and the Internet have revitalized utopian proposals by offering new platforms for their representation and dissemination, as well as the resulting critical dialogue. Characteristics of digital culture, such as access to data, speed of information, collaboration, interconnectivity across a global spectrum, victory over physical limitations, and freedom of expression, are often the principles of contemporary visual and literary utopias. The twentieth century's fascination with science fiction, Jameson argues, while utopian inasmuch as it is a response to social changes brought on by industrial progress and technological development, ultimately points up the impossibility of imagining utopia. This antinomy flows from the conflict of notions of progress and development whereby science and technology offer a harmonious, safe, and user-friendly society, but with the ever-present risk that a scientifically managed society will become an omnipresent, oppressive, and unpredictable force to be revolted against. Put another way, technology simultaneously embodies salvation and demise. Utopian projects never succeed, flawed by their failure to acknowledge the heterogeneity of the human condition. Yet this doesn't keep us from dreaming of them.

Examinations of cyberspace and the new reality, which Jean Baudrillard refers to as "hyperreality," embody the negotiations intrinsic to these cultural shifts. The social changes of recent years have injected our culture with a creative energy and revitalized productivity, but this is not without a skeptical overtone related to perceived risks and the possibility of failure. The architecture of the Internet and other technologically mediated structures of communication and representation might be considered an extension of the transparent architecture Michel Foucault identifies in the panopticon structure in his history of the penal system. Foucault argues that even when prisons were clearly identified as sites susceptible to abuses of power, society could no longer see any other alternative to making prisons "...the detestable solution, which one seems unable to do without."[7] Similarly, theorists such as Howard Rheingold warn us of the potential for the misuse of power within a technology-dominant society, and more specifically the inevitable manipulation of our understandings of need and desire and their societal articulation through the highly sophisticated, media-saturated world in which we now live.[8] To some, the technological advances infiltrating our society carry immense utopic potential, while to the neo-Luddites, their ubiquity threatens democracy as we know it. The spaces presented in *Out of Site* range from optimistic and transcendent to paranoiac and frightening, but each artist interprets the vast landscape of possibilities and challenges in which we find ourselves today.

ABSTRACT ENVIRONMENTS

In the paintings of Adam Ross, the city exists in a vast landscape of television blue. Despite a discreet frontal horizon line, buildings seem to hover in an artificial void, creating a palpable tension between the density and hyper-organization of the geometric forms that make up the cityscape and the ethereality of a site pregnant with entropy. The futuristic cities in Ross's paintings have been described as so technologically advanced that humans are no longer required, and indeed there are no human forms inhabiting his sites. Yet, one can imagine who might live in such a place. Just as the physical reality of these sites portend a resolute integration of advanced technologies into the built environment enabling flawless interconnectivity and virtual experience, those who dwell in these spaces would likely represent a hybrid of man and machine. An avid reader of science fiction novelists from Philip K. Dick to Neal Stephenson, Ross continually questions the future repercussions of our commitment to progress and advancement. If biogenetics enables us to defeat diseases and other naturally occurring human flaws, robotics may some day negate the need for the human body altogether. But what of consciousness and emotion? Will the pursuit of artificial intelligence someday supersede the uniqueness of the individual? Ross's nonfigurative works slyly raise these questions.

Ross imagines highly stylized spaces, nearly sterile in their pristine aestheticism. While the content of his painting aligns itself with contemporary science fiction, the formal qualities are grounded in the history of the utopian impulse within modern art and architecture. Taking the constructivists and avant-garde of the 1930s as a starting point, Ross works in series whose backgrounds are reminiscent of Malevich's monochromes. But rather than embracing the inevitability of completely white space, Ross imbues his painting with the saturated colors of digital culture—vermilion red, virtual studio blue, and

ADAM ROSS
UNTITLED (TOO FAR FOR THE EYE TO SEE,
ALWAYS AT THE BACK OF MY MIND #2), 2001
OIL, ALKYD, AND ACRYLIC ON CANVAS
48 X 60 INCHES
PRIVATE COLLECTION, NEW YORK

KASIMIR MALEVICH
DYNAMIC SUPREMATISM, 1915-16
OIL ON CANVAS
COLLECTION OF THE TATE GALLERY, LONDON
COPYRIGHT TATE GALLERY, LONDON/ART
RESOURCE, NEW YORK

imperial purple, sometimes incorporating horizontal sections of the hazy atmospheric colors of a post-apocalyptic or pollution-saturated industrial sky. In his titles, Ross often refers to particular influences, such as El Lissitsky, thereby linking his repeating idealized geometric shapes to the mass production and standardization of forms believed by modernist architects, particularly those associated with the Bauhaus of the 1920s, to hold the key to massive social change.

Ironically, the perpetual march of progress in Ross's worlds has led to largely static spaces, a sense of hermeticism underscored by the lack of contextualizing landscape. Despite the implication of speed of movement, in these paintings the individual elements are not visibly connected. They are stacked and layered and spatially related, yet no matrix of gridded lines permeates the spaces. Indeed, some forms float unmoored on the top half of the panels, providing respite from the gravity-bound buildings along the horizon line and conjuring thoughts of space travel or individual dwellings in a perpetual state of flux. The depiction of a city located in the non-space of what might be interpreted as open sky calls to mind some of the futuristic proposals that came out of the Russian revolution in an optimistic phase, such as Alexander Lavinsky's 1923 *City in the Air* and George Krutikov's floating apartment building of 1928. Ross's predominantly vertical and elongated shapes suggest skyscrapers, smoke stacks, configurations of housing or office buildings, even silos, yet these shapes are so fantastical in their formal simplicity as to negate direct references. Among the hard-edged objects, more painterly vertical white lines fluctuate in thickness and density as they electrify the spaces in a kind of force field. While as paintings, the works defy depth, the use of multiple perspectives in flux produces a spatial infinity. The worlds Ross creates play on our fears of loss of control and individuality in an infinite digital culture, yet the ever-visible human hand of the painter is an optimistic reminder that we control our own destiny.

In Don DeLillo's novel *White Noise*, the quiet, unremarkable, "livable" town in which the protagonist, ironically named Jack Gladney, resides, is forever changed by an "airborne toxic event":

Beneath the cloud of vaporized chemicals, the scene was one of urgency and operatic chaos. Floodlights swept across the switching yard. Army helicopters hovered at various points, shining additional lights down on the scene. Colored lights from police cruisers crisscrossed these wider beams.... Fire engines were deployed at a distance. I could hear sirens, voices calling through bullhorns, a layer of radio static causing small warps in the frosty air. Men raced from one vehicle to another, unpacked equipment, carried empty stretchers. Other men in bright Mylex suits and respirator masks moved slowly through the luminous haze, carrying death-measuring instruments.[9]

Dannielle Tegeder's underground worlds would be safe havens from just such biological or chemical catastrophes. Her paintings are color-saturated networks of interconnected systems, each element holding a distinct purpose, its ability to function reliant upon the efficacy of the surrounding parts. Seemingly abstract and deliriously decorative panels of both artificial and earth colors applied in an impressive range of techniques, Tegeder's self-sustaining cities take their visual cues from a bevy of sources—architectural blueprints, plumbing and heating systems, military reconnaissance maps, aviation routes, subway lines, disease migration maps, weather patterns, and the like. Tegeder's compositions reflect a propensity to map information through visual means, a notably resurgent impulse now that the Internet has integrated access to information with a graphic interface in which to present it. Drawing inspiration from the idealized forms of such architects as Paolo Soleri and Buckminster Fuller, Tegeder repeats particular motifs throughout her work—igloos, domes, silos, boilers, pods, cones, crystalline mounds, configurations of squares and stacked ovals, intricate grid work, complex antennae-like flagella, interconnected nodes. She has developed a vocabulary to describe each element, such as "checkered routes" and "expulsion areas," and her consideration of its function results in intricate and carefully composed spaces. No matter how fantastical her cities, like numerous utopian predecesors, Tegeder intends to create spaces that work—transportation systems lead to safety habitats, "love dot boilers" create energy for masses of electrical wiring, stacks located above ground release toxins, and so on.

There is a spontaneous painterly quality in every one of Tegeder's gestures. Each mark is carefully embellished with her imaginative flair. Particularly compelling is the use of color, an amalgamation of the life-affirming yellows, oranges, and greens seen in Indian and Persian

DANNIELLE TEGEDER
CREAM UNDER CITY WITH SECRET DOME FOREST AND ESCAPE TRANSPORT PLAN, 2001-02
ACRYLIC, ENAMEL, AND MIXED MEDIA ON PANEL
48 X 48 INCHES
COLLECTION OF EDITH AND JOSEPH DE CHIARA, NEW YORK
PHOTO: BILL ORCUTT

STEPHEN HENDEE
SUPERTHRIVE, 2000
INSTALLATION VIEW
FOAM CORE, TAPE, AND LIGHTING
COURTESY OF THE ARTIST AND RICE UNIVERSITY GALLERY, HOUSTON
PHOTO: SERGIO FERNANDEZ

PATRICK MEAGHER
INTEL OUTSIDE, 2000
DIGITAL SCULPTURE CONSTRUCTION FROM UNITBEAD
EPSON PRINT
8 X 10 INCHES
COURTESY OF THE ARTIST

MARIKO MORI
EMPTY DREAM, 1995
CIBACHROME PRINT, ALUMINUM, WOOD, AND SMOKE ALUMINUM
108 X 288 X 3 INCHES
COURTESY OF DEITCH PROJECTS, NEW YORK

miniatures, mandalas, and landscapes during her travels and the more saturated colors of New York City, found on everything from packaging to fashion to window displays. The references to love and lust in her titles—her last show was titled *Love, Lust and Other Mechanical Systems*—and the sense of optimism that the work ultimately engenders imply that utopian endeavors spring from the human impulse to create spaces where love and safety can flourish.

IMMERSIVE SITES

Stephen Hendee's environments strike a delicate balance between natural formations and digital space. His site-specific sculptural habitats are like spore growths run amok, covering a space with multiplaned crystalline forms. For *Out of Site*, these organic forms are juxtaposed with the linear organizing principle of the New Museum's building and refer to sources from the underground rock formations upon which the museum is erected to the early human dwelling of the cave or hut, or their reformulation in the "uterine architecture" developed by Tristan Tzara in the 1930s. Back-lit with primary colors, the white foam core or translucent plastic panels that are the primary material in Hendee's sites come alive in their plugged-in state. The lighting provides a harmonizing, even glow. These are not landscapes of temporal light values, brightness and shadow; instead, an artificial luminescence permeates every plane, engendering a contemplative atmosphere that is both comforting and daunting. Hendee suggests that in our post-industrial, technological world, the sublime is no longer to be found in the landscape, but rather in the solace of the computer monitor and all it contains, which we stumble into when entering his installations, much like Alice falling down the rabbit hole.

Hendee is clearly knowledgeable about the geometric growth patterns of organic forms—from crystal compounds to quartz formations—and his modular works like *Silent Sector* draw on these visual sources. Yet, his spatial understanding of the built environment is equally indebted to modernist architecture, web site design, the infrastructure of programming tools, and science fiction, wherein notions of infinite space or hyperreality are always visually articulated and reined in through the formal device of geometric forms. Hendee's work visualizes the architect Marcos Novak's descriptions of cyberspace as a "habitat

of and for the imagination" with an architecture all its own.[10] But rather than make distinctions between the realms of the natural and the artificial, Hendee opts for the hybrid, understanding that recent technological developments have blurred these categories, as witnessed in the artificial landscapes of Japanese indoor beaches and ski mountains or the more insidious possibility of posthuman robots.

UnitBead 2.0 is a spherical world made up entirely of digitized remnants of a postindustrial society configured to form an urbanscape perpetually in flux. Formally referring both to the tiny circular particles that make up Styrofoam and the infinite interstices between structures that cover the planet, Patrick Meagher's user-navigable environment keeps visitors moving through configurations in a virtual reality so versatile that no single experience of it can be reproduced. This is virtual space in all its glory—disorienting multiple perspectives, a liberating negation of gravity, boundless ebbs and flows. His "virtual tableaus," in absorbing tones of white and gray as well as saturated blues, greens, and yellows, may be flat and frontal or floating on a distant horizon line, sometimes obscured by a strange, synthetic fog. Trained in landscape architecture and urbanism, Meagher investigates the shifting spatial understandings that stem from a technology-driven urban environment. His virtual world begins with the mundane refuse of everyday life in the world of technology. Like another artist in *Out of Site*, Shirley Tse, Meagher considers Styrofoam the quintessential material of the twentieth century. Downloading styrofoam packaging material shapes from the websites of leaders in the technology industry—Macintosh, Hewlett-Packard, Intel, and Toshiba—*UnitBead 2.0* is divided into quadrants, the specific forms of the Styrofoam packaging for each of the companies' products inhabiting one section. Forms are stacked and juxtaposed to one another, individuated, and interrelated, becoming the buildings of a virtual cityscape where visitors can move around as if on an architectural tour.

Meagher is particularly interested in hybrid forms and the articulation of ideas through a wide range of media. He has described himself as a digital sculptor and virtual world photographer. Photography may be at the heart of his examinations of space, yet he creates sculpture, paintings, video, digital prints, even calendars and Styrofoam products under the rubric of ANE

Media Arts Collective, a business Meagher established in 1993 to play up the overlap between art and commerce. His prolific output includes visual representations of the interplay between technology as today's primary industry, design, architecture, and landscaping. Included in *Out of Site* are four digital captures from *UnitBead 2.0*, one from each of the quadrants. The renderings of locations within the sphere of *UnitBead 2.0* bring the well-established practice of architectural photography in the physical realm firmly into the digital. Meagher's prints can be considered examples of what Peter Lunenfeld defines as "imagescapes" or electronic facades, which, he argues, challenge our very concept of architecture, an impulse this artist most certainly shares.[11] Meagher demonstrates a commitment to new forms of visual expression by neologizing—terms like "blandscapes," "styroforms," "styroportraits," and "photage" describe his varied output—a practice apparent in cyberculture, where the language of critical discourse can barely keep pace with changes in reality.

Haluk Akakçe's double digital projection piece *Still Life* begins with the image of a large camera on one screen and a rainbow of vertical images on the other, like the test screen before a television show begins. As the colors' edges begin to blur into a more painterly, seductive surface, the camera rotates and suddenly we enter the lens into a new world. Both screens then simultaneously provide an exterior and interior view of an urban corporate office. Situated against a skyline of buildings, the office contains an empty chair facing the window. A matrix of gridded lines becomes the formal device for numerous mutations, reflecting both the organizing principle of the modern cityscape and the digital landscape in which all of Akakçe imagery is built. Shapes morph and perspectives shift, each time bringing us to a familiar, yet strangely disturbing site. We enter a room completely empty save for three discreet things, each moving on an internal axis: an ornately decorative object turns like a child's top; the figure of a man moves his arms futilely in a circular swimmer's motion; and a miniature house with a mirrored exterior rotates, picking up reflections of the surfaces of the room. Each thing seems to have a distinct purpose, which eludes simple interpretation. Like a lab filled with idiosyncratic experiments, the room carries a heavy

weight of industry and technological invention but remains a realm of everyday business with its modest scale and domestic decor. This is a psychologically charged place where the business of electronica dominates and isolation reigns, like living inside the television Akakçe alludes to at the beginning of the piece. But rather than presenting a pure dystopia completely skeptical of technological advancements and their homogenizing tendencies, Akakçe has us enter the rotating house to find ourselves in a calm autumn environment of trees and fallen leaves. Space morphs again so that we remain outside the house and watch it suspended and rotating in the air, while the vast world around it is reflected upon its surface.

The mirrors in Akakçe's work simulate the surfaces of the modern urban skyscraper, yet depicted on a domestic site which ultimately finds itself resituated in nature, the mirror becomes a perceptual plane functioning to create a hybrid space of architecture and environment, interior and exterior. Anthony Vidler argues that contemporary ideologies of the city and notions of monumentality call for a return to transparency that is grounded in an uneasiness and a desire to "disappear, be invisible."[12] He identifies the history of the mirror's relationship to the uncanny—wherein distinctions between inside and outside, self and others become indecipherable. Akakçe takes us on an architectural journey replete with the anxieties of urban space and a tempo of continual transformations and mutations, visualizing the experience Vidler describes when he writes, "...the uncanny itself is framed as a sudden apparition seen, as it were, through a window."[13]

THE PHOTOGRAPHIC UNCANNY

The blurry boundary between real and virtual space is the site Craig Kalpakjian's photographs explore. His corporate office building interiors are visually familiar enough for the viewer to assume they are photographs of existing sites. But closer examination reveals that these hermetic, pristine spaces of dry wall, drop ceilings, mirrors, and linoleum floors are slightly askew. Kalpakjian is careful to include details such as perfectly spaced light fixtures, heating and air ducts, and other markers of a building's infrastructure. Although the viewer may fail to notice initially, he also provides visual clues that the

CRAIG KALPAKJIAN
STAIR, 2001
CIBACHROME PRINT ON ALUMINUM
30 1/2 X 58 1/2 INCHES
COURTESY OF ANDREA ROSEN GALLERY, NEW YORK

sites are completely fabricated. A long hallway contains not a single door; a lobby, rather than providing access to elevators and stairs, has only one tightly shut door. In many of the images, light beckons us forward but a sinister quality saturates the work, turning it into something like a bad dream in which each corridor leads to its mirror image, a vortex of lost space. Using such tools as the programs Form Z and Lightscape now available to architects and designers, Kalpakjian's uncanny spaces are digitally rendered and later produced as cibachrome prints, a medium more typically associated with architectural photography. Virtual space is presented here as something tangible, extending the argument that photography's role as a documentary medium has been obliterated in the digital age.

Kalpakjian's depictions of particular details in these interiors reflect an interest in tools of surveillance and control that have become so commonplace in public and private sites as to barely cause a blip on our visual radar—security mirrors in corners, speakers installed flush with the ceiling, surveillance cameras pointing at doorways. By zeroing in on these, Kalpakjian underscores our propensity to accept them in our peripheral vision and asks whether our fear of crime and litigiousness has tipped the scales too far toward militarism and caused us to give up our individual freedoms in the process. By focusing on the private sites of business, he pushes this narrative into the realm of corporate espionage and efforts to prevent it, creating locales that in their visual banality subvert the suspicion that your every move is being monitored. As Anthony Vidler has observed, individual sites and their spatial configurations influence psychological states to such a degree that communities have historically attributed specific mental disorders, from claustrophobia to anxiety, to trends in urban and suburban growth. In his discussion of one type of "spatial warping," Vidler writes:

The first is that produced by the psychological culture of modernism from the late nineteenth century to the present, with its emphasis on the nature of space as a projection of the subject, and thus as a harbinger and repository of all the neuroses and phobias of that subject. Space, in this inscription, is not empty, but full of disturbing objects and forms, among which the forms of architecture and the city take their place.[14]

Kalpakjian captures the psychology of urban sites and asks what social malaise will result from their presence.

The conflation of "real" experience with the "virtual" is breaking down conventional distinctions between the physical body and the space it inhabits. In *Megalopolis: Contemporary Cultural Sensibilities*, the cultural critic Celeste Olalquiaga argues that contemporary experience causes the body, as an individual entity or identity, to abandon itself in an effort to understand the space that surrounds it, resulting in an inability to differentiate between the two, a process she describes as being "swallowed" by the vastness of the world.[15] The *Interior* series by the collaborative team Anthony Aziz and Sammy Cucher takes contemporary disorienting experiences of space, in which the distinction between exterior and interior collapse, to a truly disturbing degree, producing a hybrid of the human body and architectural sites.

Interior is a series of digital photographs of stairwells, corridors, and confined spaces covered with human skin. The geometric precision, subdued lighting, and familiar corporeal tones of these spaces is initially seductive. But this comforting illusion is quickly undermined as one becomes aware of the visible characteristics and flaws of the skin—freckles, moles, pores, hairs, and pimples—on the wall, floor, and ceiling surfaces of the space. In its creepy projection of what the future of biotechnology may hold, Aziz + Cucher's imagery aligns itself, in part, with science fiction's premonitory role. More importantly, the works function as a metaphor for the body's relationship with space in the digital age. Marcos Novak describes cyberspace as the quintessential site in which traditional distinctions between the mind and body break down. As a result, reality is increasingly devised by our imagination, so that "while we reassert the body, we grant it the freedom to change at a whim, to become liquid."[16] In this sense, the "liquid architecture" of Novak's thesis would seem to find its visual equivalent in Aziz + Cucher's interiors where the body actually morphs into the site it inhabits.

SKEWED PERSPECTIVES

Whereas the social, and more specifically, spatial impact of digital technology informs the conceptual underpinnings of all the painters in *Out of Site*, several of them use little or no digital technique in their processes. Kevin Zucker directly incorporates the digital into the more conventional application of paint on canvas, making paintings that interface the spatial considerations of digital experiences with the two-dimensional framework of the canvas. Part of the complex process of creating these elegant, ghostly interiors is downloading images of elements of vernacular spaces from the Internet, printing them with ink-jet, and transferring them to the canvas. Only in his very recent work, such as *World Without End*, has Zucker moved beyond using simple drawing programs to more advanced software—in this case, to work out the infinite space resulting from the reflections of two adjacent mirrors in a dressing room. Zucker's mediation of imagery functions to critique the austerity and presumed autonomy of modernist painting and architecture. The gridded floors and ceilings ultimately defy seriality and a rational organization of space by becoming the foundations of awkward perspectives.

Absurdly baroque objects invade Zucker's pristine environments of washed out, muted gray and brown tones. As Peter Lunenfeld argues in his discussion of technology's impact upon architecture, as opposed to the use of technology by the modernist movement to create a clean aesthetic, the multifarious attributes of today's technology portend a future that embraces "ornamentation" and "mutable imagescapes."[17] In Zucker's *Angels: The Heads of Pins*, for example, a gigantic chandelier dominates an empty ballroom, dangling precariously from a single anchor in the ceiling. The excessive detailing and decoration of this object make for a weighty counterpoint to the desolate context. One yearns to know more about these interiors, so familiar in their embrace of the everyday and yet fantastic. These are the dead institutional spaces that have invaded public life—from hotels to schools to airports—desperately trying to distract us from their oppressive nature through their seductive surfaces of crystals, mirrors, and velvet rope. Amidst the precision and elegance of the image, countless surface imperfections exist. In the process of transferring the printed imagery to the canvas, tears and rips occur. This loss of information is inevitable in any process of translation—in language or in representations found in the media or visual arts—and is particularly poignant in the post-industrial "information age" where every endeavor is wholly reliant upon access to and speed of data. Yet,

KEVIN ZUCKER
WORLD WITHOUT END, 2001
ACRYLIC, INK-JET AND CARBON TRANSFERS, AND ENAMEL ON CANVAS
56 X 85 INCHES
COLLECTION OF DOUGLAS S. CRAMER, ROXBURY, CT
COURTESY OF MARY BOONE GALLERY, NEW YORK

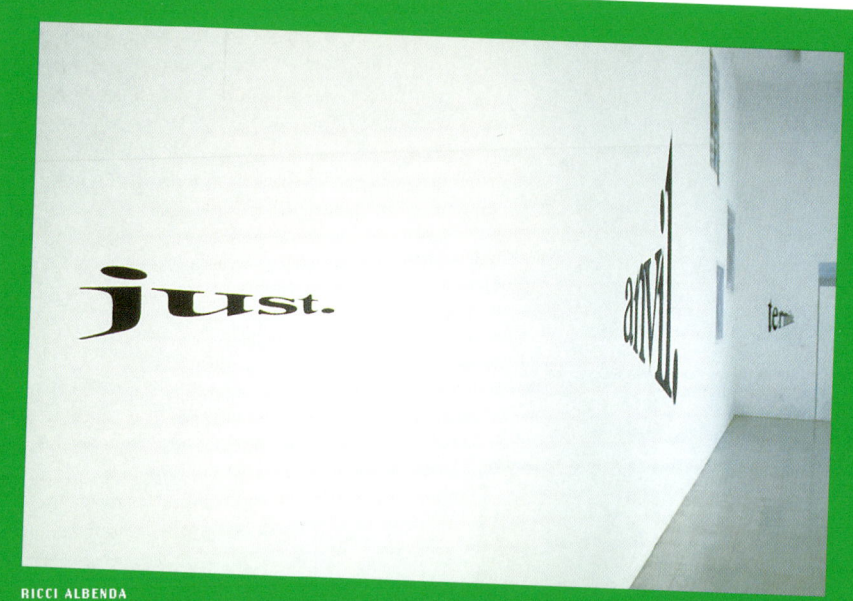

RICCI ALBENDA
JUST, 1996
ACRYLIC ON WALL
5 X 15 INCHES
ANVIL, 1996
ACRYLIC ON WALL
12 X 12 INCHES
INSTALLATION VIEW: ARLINGTON ART MUSEUM, TEXAS
COURTESY OF ANDREW KREPS GALLERY, NEW YORK

CANNON HUDSON
COLOSSUS 1, 2002
OIL ON CANVAS
72 X 54 INCHES
COURTESY OF THE ARTIST
PHOTO: JASON MANDELLA

Zucker suggests, our impatience for knowledge and quick answers means that things get overlooked and lost in the network.

Thick green bands run vertically down a background wall and then suddenly turn horizontally, beginning to lose their composure as they move across the interior space depicted. In *Colossus I* and *Colossus II*, blue and red self-contained water-like flows begin arbitrarily at the edge of a pool, run across slate gray tiles, and then dramatically drop to the bottom of the picture plane. *Serpent* is full of multiple planes vying for space in a compressed room. A shiny, industrial looking object pushes in from the right and hovers in a pink void, a kind of hybrid of furniture, the infrastructure of built environments, and inflatable toys. These are the strangely inviting interiors of Cannon Hudson.

Focusing on small areas of apparently much larger spaces—corners, part of a floor and wall, a section of a tub or pool—Hudson's imagery is always foregrounded to such an extent that we sense we could never experience a physical space from such a simultaneously intimate and open viewpoint. His paintings are full of disconcertingly skewed perspectives. Stripes of color reminiscent of 1970s interior design textiles often move through the interiors in a series of twists and turns, creating depth by juxtaposing this spatial extreme to the more architectural elements of the work. The earth-toned field in the top corner of *Colossus II* creates the illusion of a window, a rare detail in the paintings which acknowledges an exterior world. Even with some push and pull between interior and exterior, Hudson continually compresses his spaces and pushes us into a corner, only to back off a bit as our fascination turns into discomfort. An interior designer himself, his works respond to interior and industrial design and modernist architecture. The rooms depicted begin with the modular rationalism of modernism and are punctuated by the detailing of tiles, brick, and nearly baroque textiles straight out of *Metropolitan Home* magazine. Hudson's spaces aren't so much surreal as they are the products of a digital age, where multiple co-existing planes, warped perspectives, and continually shifting surfaces are embraced with revolutionary fervor.

Ricci Albenda creates physical manifestations of theoretical spaces, from simple perspectival manipulations of rectilinear architectural interiors to complex visualizations of multi-dimensional perceived forms. Despite its grounding in historical investigations of perception and complex mathematical equations, one's primary experience of Albenda's work is visceral. His site-specific installations provide a personal and distinct understanding of the body's relationship to space, which shifts and mutates depending on your position in the gallery. Quiet, sophisticated, even austere, the works engender a continual sense of motion. To experience a work such as *Tesseract*, a room-size investigation of the four-dimensional hyper-cube, is to be dropped into a realm of infinite space. The infinity of space has been imagined by numerous artists, but Albenda is able to place the viewer physically within these fantastic notions, making possible a profoundly tangible experience of the virtual. Space is sculptural for Albenda, something that can be grabbed, manipulated, and re-presented as altered reality.

Albenda curves existing walls in convex forms so that they begin to pulse and breathe quietly. He rearticulates the height and horizontal line of walls by painting slightly skewed bands of nondescript gray at the top of the space. More conspicuously, he inserts biomorphic "portals" flush with the wall—beautifully made vortexes designed to transport us into another dimension. Albenda questions what we know about space and responds to the renewed need for this investigation in the digital age. His text pieces use existing architecture to play up the varying vantage points of the gallery. Executed in vinyl letters adhered directly to the wall, from one perspective the chosen word may be elongated and illegible, as if viewed through a warped lens. Take a walk through the gallery and that same word is suddenly crystal clear, its meaning shifting just as quickly as your perspective. Albenda's choice of words in past text pieces—"small," "single," "just," "algae," "breathe," "yellow"—can be interpreted as relating directly to space, but are often enigmatic. What's important here is not exclusively the individual words but the way they illustrate how meaning is wrapped up in perspective.

For *Out of Site*, Albenda pushes his text work in a new direction. The word "people"—all lowercase in the elegant font developed by the artist—is produced in numerous sizes and covers one wall of the gallery. Interacting directly with the architecture of the gallery, the potential homogeneity of the formal quality of the words is undermined by the surprising amount of difference revealed within the shifting scales and perspectives. Referencing the urban experience of numerous people in geographically and architecturally small spaces, this piece speaks to the realities of navigating the cityscape while remaining a more abstract investigation of the body's relationship to space. With poetic resonance, in all of his work, Albenda is able to take a theoretical notion and construct it physically so that it carries phenomenological weight.

TOPOGRAPHIES AND NETWORKS

Nina Bovasso's acrylic on paper drawings are focused environments of built space—as meandering, temporal, precise, structured, or as risky and precarious as any contemporary metropolis. In irresistibly decorative conglomerations of shapes and lines, she painstakingly stacks and extends squares, rectangles, circles, dots, arches, cones, towers, blobs, even hearts, each form relating to its surrounding supporting structure until its mass inhabits the picture frame just enough to seem simultaneously substantial and ephemeral. She embraces abstraction's particular acuity in investigating space and through a dizzying repetition of motifs, keeps the relationship of figure to ground a central component of the work. Although nearly uncontrollably pushing out in all directions, the imagery emphasizes a vertical movement. The bottom edge of the paper becomes the ground from which Bovasso's absurdist cities begin to take up space.

Cities around the world are growing at phenomenal rates, perpetually changing and unstable. Responding to the impact of urban growth on our understanding of spatial organization, Bovasso's abstractions are immediately recognizable as hypertrophic urban sites, struggling against self-annihilation. All of her drawings seem to embody a balancing act, like a house of cards where the addition of one more object may cause the whole structure to topple to the ground. Delightfully lyrical and formally coherent, the metonymic structure of her drawing process captures the complexity of the city where each individual structure represents this highly organized, yet somehow chaotic whole many of us call home.

Matthew Northridge's version of urban sprawl, *New City*, is rendered sculpturally, reminiscent of an architectural model. Using more than three thousand individual pieces of variously sized rectangular forms cut out of Masonite and meticulously covered with imagery found in available printed materials, the elements are placed side by side and stacked onto a large, square platform, providing a glimpse into a fantastical, Lilliputian city which one imagines could go on forever. Northridge's city is playful in its toy-like configuration yet relentless in its territorializing impulse. Only two feet high, the viewer hovers over it with a pronounced aerial view. *New City* leaves no room for expanses of horizontal landscape or even the smallest reminder of the so-called natural. Rather, all available space is gobbled up by vertical growth—a dense conglomeration of buildings, each one individualized yet so modulated as to become interchangeable.

Northridge begins with available information. Taking imagery from popular printed sources—magazines, books, newspapers, packaging—he is interested in the infinite possibilities the materials of everyday life provide. Cutting out sections from these sources, he alters the integrity of the information as it was originally dispersed, yet it remains recognizable as culturally specific. Urban experience is marked by the overwhelming amount of visual information thrust upon us as we navigate its spaces. Storefront awnings, billboards, subway signs, neon pronouncements, window displays—in the melee of urban life, the specific can get lost amidst the vastness of input out there. Similarly, the surfaces of Northridge's buildings are replete with visual information that in its expansiveness and disjointedness has lost definition, like walking through Times Square. To gain insight into a specific reference, we must slow down and focus in. In *New City*, the architectural forms are inseparable from the sources of information. Despite its tactile quality and the amount of hands-on work required to make this piece, this is a city of the future where information and architecture are seamless.

Suburban growth has been on the rise ever since the developer William Levitt capitalized on the economic boom and demand for single-family housing following World War II. Population growth coupled with the problems endemic to the modern city, such as crime, overcrowding, and pollution, made suburban life appealing, and the suburbs continue to be touted as safe, clean, and convenient, however false these claims have turned out to be. Throughout the world, natural

NINA BOVASSO
UNTITLED, 1999
ACRYLIC ON PAPER
27 1/2 X 40 INCHES
COLLECTION OF DEREK ELLER & ABBY MESSITTE, BROOKLYN

SVEN PÅHLSSON
WITH COMPOSER ERIK WOLLO
SPRAWLVILLE OR LIFE AT HIGHWAY EXIT RAMP, 2001
3D ANIMATION WITH SOUND
10 MINUTES
COURTESY OF SPENCER BROWNSTONE GALLERY, NEW YORK

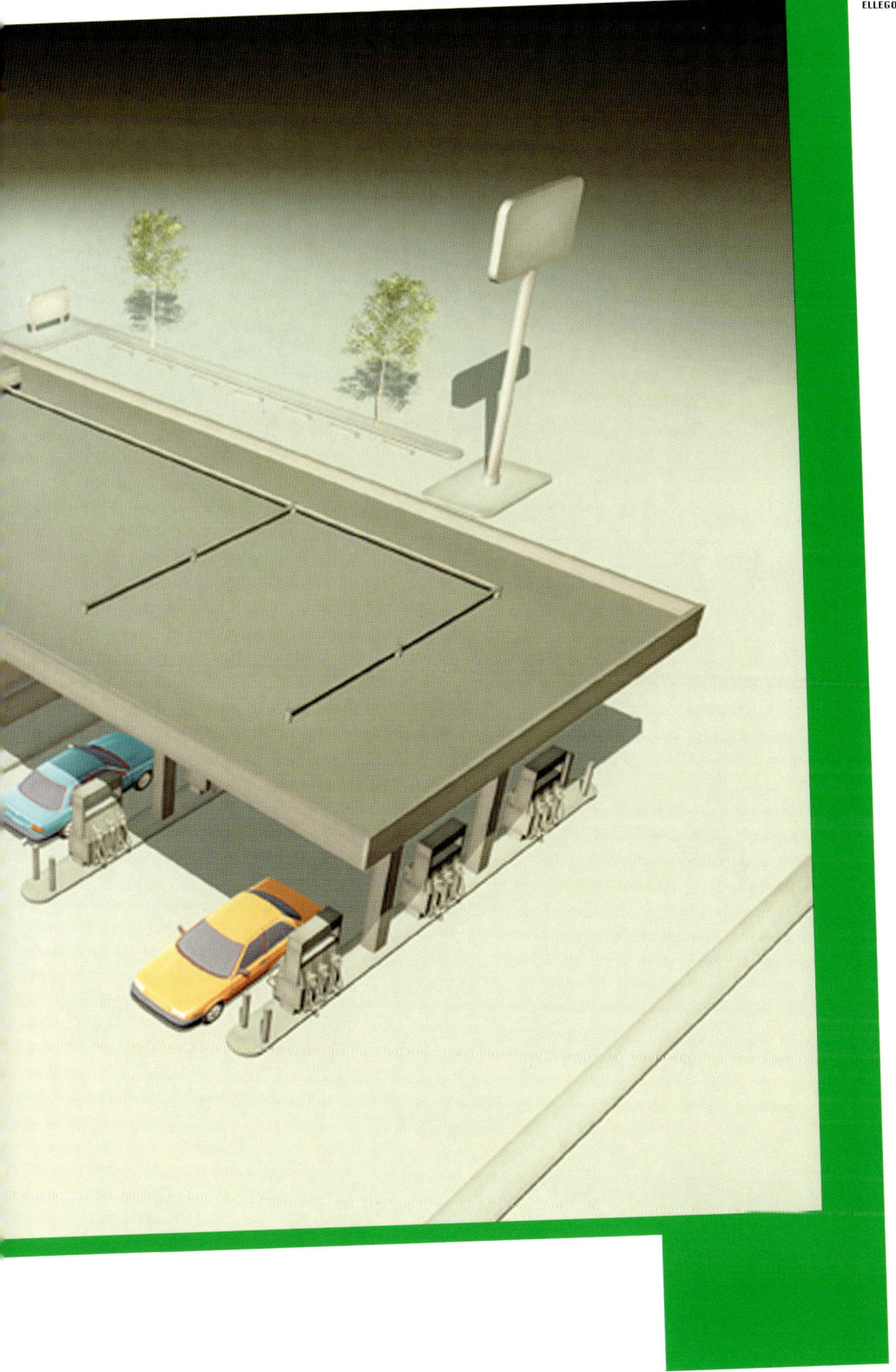

habitats have given way to the architectural structures erected to house suburban dwellers. Suburban growth is also emblematic of the desirability of a sense of progress. Without the standardization and modularization of building or the construction of highways, these communities would not be possible. In his digital 3D animation, Sven Påhlsson takes us on a journey through a fictional suburban landscape bitingly dubbed *Sprawlville*. A Norwegian who splits his time between Oslo and New York City, Påhlsson has spent a great deal of time in his car traveling the interstates of the United States. Although one won't find *Sprawlville* on a map, it is informed by Påhlsson's road trips and could function as a prototype for the development currently spanning the globe. The physical markers of suburban lifestyle in the work are immediately recognizable to the viewer—homogeneous housing built side-by-side, expanses of strip malls located along roadways calling forth with neon lights, massive shopping malls surrounded by countless parking spaces, and a complex system of paved roads connecting it all together.

To create *Sprawlville*, Påhlsson used digital modeling programs, which he relates to the stream-lined building processes of suburban growth, where repetition of form is so acceptable as to be considered desirable. He captures the general anxiety of the suburbs, where all the optimism they offered the inhabitants is gradually overshadowed by the reality of their failed utopic aspirations. The demand for a safe haven has resulted in development imprisoned by its own fear, where walls and gates, high-tech surveillance systems, fenced yards, and tightly locked windows and doors keep people isolated inside rather than encouraging them to be part of an active community. Påhlsson takes us on a circuitous trip through *Sprawlville*, from aerial views to close-ups of parking lots filled with shopping carts and highways with cars zooming by on either side. The high resolution of the animation captures the contradictory essence of suburbia, for it is visibly artificial and non-descript and yet enlivened by the constant motion, bright colors, and enticing lights of a space in which human activity is teeming. Påhlsson's aerial views of *Sprawlville* connect his choice of medium concisely to the technology that enables it. The infrastructure of gridded neighborhood streets and parking lots, interconnected via a complex system of roads,

calls to mind a digital matrix allowing for the continual flow of massive amounts of data. Påhlsson's digital architectural renderings make apparent that the same methodology of building that takes place in the physical realm happens in virtual reality.

Like Stephen Hendee, Shirley Tse's sculptures suggest that the natural and the artificial are not mutually exclusive. Consisting of intricate shapes carved into the surfaces of sections of sky blue sheets of polystyrene, *Polymathicstyrene* is a meandering topography installed waist-high like a shelf along the gallery's walls. Navigating the space of the gallery in order to take in the work's nuances, visitors have an aerial view of a horizontal core sample of landscapes and cityscapes shifting and morphing over time. In Tse's seemingly abstract sculpture, which uses both hard-edged geometric forms and fluid lines, countless familiar architectural and environmental sites emerge from the clean, monochromatic surface of the plastic—stairs, mazes, buildings, arenas, reservoirs, rivers, pyramids, hills, rock formations. Each section is inspired by different elements of our culture—some are aligned with technology and include forms resembling computer monitors, data chips, and control panels; others are other-worldly, like the surface of another planet or a spaceship launching pad. Despite the eye's tendency to articulate particular areas in familiar terms, Tse turns conventional diametric oppositions such as indoor/outdoor and artificial/natural on their head. Her sites are beautiful yet fake, referential yet fantastic, visually cohesive yet infinitely variable.

Tse's practice embraces a near fetishism about plastic. The most ubiquitous material of the global trade economy, plastic packaging has a contradictory role. From Styrofoam to bubble wrap, it is valued for its adaptability and variability in protecting a product during shipping; then it is discarded as useless debris once a product reaches its destination and derided as a threat to the environment because it lacks biodegradability. Growing up in Hong Kong, Tse was fascinated by the stacked metal containers in the city's shipping yards and realized early on that the primary function of society is to assure the constant flow of products to people. In a transient world, where malleability and flexibility ensure survival, plastic for Tse is the quintessential symbol of modern life. In *Bionicpak*, she begins with two identical

SHIRLEY TSE
POLYMATHICSTYRENE, 2000
EXTRUDED POLYSTYRENE
DIMENSIONS VARIABLE
COURTESY OF MURRAY GUY GALLERY, NEW YORK, AND SHOSHANA WAYNE GALLERY, SANTA MONICA

JULIE MEHRETU
RETOPISTICS: A RENEGADE EXCAVATION (DETAIL), 2001
INK AND ACRYLIC ON CANVAS
96 X 216 INCHES
DIMITRI DASKOLOPOULOS COLLECTION, ATHENS
COURTESY OF THE PROJECT, NEW YORK AND LOS ANGELES

FRANK LLOYD WRIGHT
BROADACRE CITY, FROM THE LIVING CITY
PUBLISHED BY HORIZON PRESS, NEW YORK, 1958
COLLECTION OF THE NEW YORK PUBLIC LIBRARY, MIRIAM AND IRA D. WALLACH DIVISION OF ART, PRINTS, AND PHOTOGRAPHS
© 2002 FRANK LLOYD WRIGHT FOUNDATION, SCOTTSDALE, AZ/ARTISTS RIGHTS SOCIETY (ARS), NEW YORK

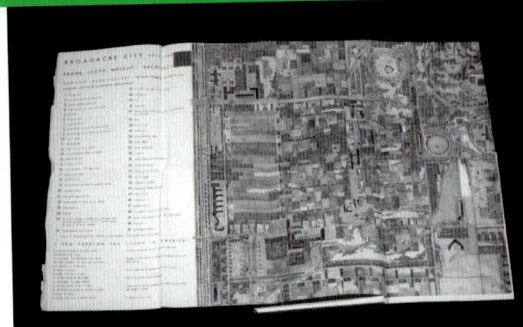

existing packing material forms and transforms them into related but distinct topographies, one lyrical and quirky, and turned on its side, the other upright, filled with the precise, geometric, modular forms of modernist architecture. Tse takes the discarded materials produced to fill the negative space in packaging and transforms them into sites to be reckoned with, much as the colonization of cyberspace creates something out of nothing. But, these are not utopian or futuristic sites. Rather, they are forms that represent the society in which we already live, where industry, landscape, technology, and humanity constantly vie for space and increasingly overlap, becoming spatial hybrids.

Julie Mehretu is part archeologist and part urban planner, using painting to tease out the vastness of spatial experience embodied in the past and present, as well as suggesting possibilities for the future. Each of her paintings is an astonishing amalgamation of disparate parts moving across the canvas like a determined processional through time and space. Layered, morphing, and fluid forms and staccato ink lines are juxtaposed and superimposed over bands of geometric colors and outlines of readily identifiable architectural structures, all accentuated and energized into a near frenzy by explosive blasts and depictions of atmospheric occurrences, from delicate cloud formations to hurricane winds. Some of the visual sources of Mehretu's topographies are quotidian and, running the gamut from abstracted to representational, are ingrained in the urban landscape: airport terminal maps, architectural floor plans, site sketches, newspaper clippings of natural disasters, subway and highway systems. But there are also references to history—the classical order of columns and arenas, and what appear to be architectural ruins, eroded over time but resilient and stubbornly physical.

A definite tension between old and new exists here, and a collapsing of time coincides with the folding in of planar fields and perspectives that keep viewers in the perpetual motion of the civilizations imbedded in Mehretu's imagination. We view all this simultaneously from numerous vantage points—aerial views of conglomerations of housing right next to a foregrounded outline of a roadway moving through space in one-point perspective, a coherent floor plan of a particular site serving as the backdrop to numerous visual

markers of human activity. Perspectives dissolve and spaces oscillate in overlapping fields of dense information with multiple vanishing points.

Mehretu's sites are indebted to science fiction and technology, suprematist utopian gestures played out through geometric abstraction, the dynamism of the futurists embrace of mechanical forms, and the cityscape drawings of such optimistically inventive theorists and practitioners as the international situationist Constant Nieuwenhuy and the Russian architect Iakov Chernikhov. Mehretu's inclusion of structures in decay—billowing smoke spirals and explosions of fire—are not harbingers of demise. Rather, they place her ideologically in the optimistic camp of artists and architects who believed that true social change could only come through total transformation, even if it meant a revolutionary destruction of the existing cultural infrastructure. These include the aggressive call to "wreck" the city in the first futurist manifesto of 1909 to Frank Lloyd Wright's proposal for *The Living City* that would completely reconfigure the urban landscape of the metropolis to encourage individuality and freedom of thought.

Mehretu's meandering images are maps of a neo-nomadic culture—a society characterized by movement, a breakdown of old borders, and an implosion of time and space. Discussing technology's impact upon the city, the architect and cultural critic Paul Virilio writes:

In this new perspective devoid of horizon, the city was entered not through a gate nor through an arc de triomphe, but rather through an electronic audience system. Users of the road were no longer understood to be inhabitants or privileged residents. They were now interlocutors in permanent transit. From this moment on, continuity no longer breaks down in space....From here, continuity is ruptured in time, in a time that advanced technologies and industrial redeployment incessantly arrange through a series of interruptions....[18]

Not surprisingly, Mehretu's personal background is notably diverse. Born in Addis Ababa, Ethiopia, she spent her childhood in Africa and Kalamazoo, Michigan, and her ancestors hail from Ethiopia, Poland, France, England, and the United States. Described by some as contemporary historical paintings, Mehretu's work, however abstract, is deeply informed by the narrative of a society in flux and a postmodern comfort with pastiche wherein the present is a hybridization of past forms reconfigured into a compellingly messy yet cohesive whole. Consider her work included in *Out of Site* titled *Retopistics: A Renegade Excavation*: it begins with a neologism that implies a re-articulation of a topography. Her map-like imagery is then identified as an unconventional uncovering of available cultural materials. The artist honors the past by acknowledging its role in the present, heeding Walter Benjamin's warning: "Every image of the past that is not recognized by the present as one of its own concerns threatens to disappear irretrievably."[19] As historical paintings of a fictional universe, replete with implied anecdotes of a society adapting to perpetual change, Mehretu's works capture this particular cultural moment.

The proliferation of digital technologies giving rise to tangible experiences of the virtual coupled with an increasing embrace of a nomadic lifestyle wherein navigations through the interstitial "no-man's lands" of airports, train stations, and border crossings as well as the ever-expanding urban centers of the world are cultural phenomena ripe for investigation by artists. A periodic reconsideration of the notion of "site" in recent years has put forth the controversial claim that the primacy of the local is being replaced by a global economic system that promotes and perpetuates homogenization. The impulse of the artists in *Out of Site* to create fictional sites responds in part to this threat. Moreover, their unique representations of sites both informed by history and emblematic of everyday experiences of space serve to relocalize the notion of site in a definitive way.

Skeptical that "the choice to fictionalize" provides permission to valorize a culture of flux and sitelessness, art and architecture historian Miwon Kwon asks whether art "might mean finding a terrain between mobilization and specificity—to be out of place with punctuality and precision."[20] The works in *Out of Site* do just that by proposing new sites that challenge the dialectics of real and fictional, utopic and dystopic, local and global, interior and exterior. In so doing they visualize for us the vast landscape of spatial possibilities that underscores the world today.

1. William Gibson, *Neuromancer* (New York: Ace Books, 1984), 5.
2. Ibid, 51.
3. Elizabeth Grosz, *Architecture from the Outside: Essays on Virtual and Real Space* (Cambridge, Mass.: The MIT Press, 2001), 75.
4. Iain Chambers, *Popular Culture: The Metropolitan Experience* (New York, London: Methuen, 1986).
5. See Anthony Vidler, *Warped Space: Art, Architecture, and Anxiety in Modern Culture* (Cambridge, Mass.: The MIT Press, 2000).
6. Fredric Jameson, "Progress Versus Utopia; or, Can We Imagine the Future?" *Science-Fiction Studies 9*, no. 2 (July 1982): 147–158, and reprinted in Brian Wallis, editor, *Art After Modernism: Rethinking Representation* (New York: New Museum of Contemporary Art, 1984), 239–252.
7. Foucault, "Complete and Austere Institutions," from *Foucault Reader*, edited by Paul Rabinow (New York: Pantheon Books, 1984), 215. See also *Foucault, Discipline and Punish: The Birth of the Prison*, translated by Alan Sheridan (New York: Pantheon Books, 1977).
8. See Howard Rheingold, *The Virtual Community: Homesteading on the Electronic Frontier* (Cambridge, Mass.: The MIT Press, 2000) and *Virtual Reality* (New York: Summit Books, 1991).
9. Don DeLillo, *White Noise* (New York: Penguin Books, 1984), 115–116.
10. Marcos Novak, "Liquid Architectures in Cyberspace," in Randall Packer and Ken Jordan, eds., *Multi-media: From Wagner to Virtual Reality* (New York: W.W. Norton & Co., 2001), 253.
11. See Peter Lunenfeld, *Snap to Grid: A User's Guide to Digital Arts, Media, and Cultures* (Cambridge, Mass.: The MIT Press, 2000).
12. Anthony Vidler, *The Architectural Uncanny: Essays in the Modern Unhomely* (Cambridge, Mass.: The MIT Press, 1992), 220.
13. Ibid, 224.
14. Vidler, *Warped Space: Art, Architecture, and Anxiety in Modern Culture*, vii.
15. Celeste Olalquiaga, *Megalopolis: Contemporary Cultural Sensibilities* (Minneapolis: University of Minnesota Press, 1992), 2.
16. Novak, 255.
17. Lunenfeld, 103.
18. Paul Virilio, *The Lost Dimension* (New York: Semiotext(e), 1991), 11, first published as *L'espace Critique* (Paris: Christian Bourgois, 1984).
19. Walter Benjamin, *Illuminations* (New York: Harcourt, Brace & World, 1968).
20. Miwon Kwon, "One Place After Another: Notes on Site-Specificity," *October* 80 (Spring 1997): 109–110. Kwon's discussion here is specifically related to "site-specific" art, and although many works in the exhibition would fall outside the convential definition of this term, much of her discussion seems applicable.

AZIZ + CUCHER
INTERIOR #7, 1998-99
C-PRINT, EDITION OF 3
62 X 40 INCHES
COURTESY OF HENRY URBACH ARCHITECTURE, NEW YORK

ADAM ROSS
UNTITLED (TOO FAR FOR THE EYE TO SEE, ALWAYS AT THE BACK OF MY MIND #3), 2001
OIL, ALKYD, AND ACRYLIC ON CANVAS
48 X 60 INCHES
COURTESY OF THE ARTIST AND SARA MELTZER GALLERY, NEW YORK

DANNIELLE TEGEDER
PINK CITY STATION WITH DEAD-END TUNNELS AND LUST FIRE, 2001
COLORED PENCIL, INK, ACRYLIC, HOUSE PAINT, DYE, PIGMENT, AND ENAMEL ON CANVAS
48 X 48 INCHES
COLLECTION SUSAN EVANS, NEW YORK
PHOTO: JASON MANDELLA

Faced with a group of emerging artists who are developing fictional architectures in response to the increasingly digitized world, it may be worth monitoring what the equivalent generation of architects have been up to.

Architecture has learned to fly again. Countless projects of the last decade have released themselves from the constraints of the ground to drift through the endless void of cyberspace, continuously morphing their form in response to diverse streams of information. Traditional drawings and models have given way to animations, scans, and readouts. Glowing translucent forms adjust their shape in front of us like biotechnological creatures against a background usually rendered as black as outer space. They twist and turn in cinematic sequences overlaid with slogans, statistics, vectors, labels, and commentary—all in the latest typefaces. The conventional understanding of a physical site has been replaced by the idea of a complexly dynamic force field. Architects solemnly cite passages from the latest thinking about biogenetics, nano-technology, cryptology, marketing, and military strategy. The long history of their own field is just an embarrassing episode that has to be ignored. Experimental architecture has become a branch of science fiction. Designers have pushed their new tools and ideas to the limit, fantasizing about an idealized future rather than the messiness of everyday life on the ground.

Architectural publications increasingly have the look of sci-fi catalogues, with schemes for virtual houses lined up alongside biomorphic figures generated from the analysis of diverse forms of traffic flow. Yet the work is not presented as science fiction. On the contrary, it is earnestly delivered as science fact, the results of precise laboratory experiments rather than speculative creations by artists. These architects don't even want to be seen as designers. The preferred role is that of the researcher, technician, or analyst. The latest software is used to give a tangible sense of realism to the representations— the computer being both the laboratory in which the experiments are carried out and the means for making the projects appear to have already been realized. The promise of virtual environments is to remove this difference between imagined and actual construction, but for now almost nothing of this work has actually been built outside the spaces of exhibitions, classrooms, lecture halls, magazines, and the Internet. On the few occasions that it is, it tends to look like a crash-landed spaceship, a deep space creature suffering under the pressure of gravity, or a jellyfish dying on a beach. All the animated life of the new forms of architectural organism being bred on the computer screen gets frozen. The need to provide a stable structure and program, traditional missions for the architect, become harsh compromising forces. This architecture is happiest off the ground in digital space.

All this echoes a similar liftoff in the 1960s when architectural discourse was deluged with images of new kinds of urban structure. Cities started to float, dive underwater, bury themselves in the desert, rise up from ice caps, suspend themselves, walk, fly, orbit the earth, or drift through outer space. They grew to the size of the planet like colossal amoeba or were compacted into efficient packages that could be launched into space. Architecture left its traditional site behind. Fixed buildings gave way to capsules, pods, webs, nets, tubes, and bubbles. Structures were suspended, plugged in, clipped on, hooked up, or inflated. And the ultimate virtues were portability, malleability, and expendability. The building as conventionally understood was systematically dissolved and abandoned.

Architecture had to drastically reduce its weight so that it could take off—throwing off most of its material and historical baggage. Only a basic scaffolding and set of equipment remained that established the condition for the physical survival of the human species and then provided for intellectual and emotional needs by enveloping people in dense layers of information and imagery. Buildings turned into vehicles, transporting their occupants both by moving through space and by continually generating new spaces.

This trajectory of projects culminated in environmental simulators, compact machines like Archigram's *Audio-Visual Jukebox* that could produce an ever-changing array of architectural effects on demand, or vast landscapes treated as playgrounds for endlessly ecstatically rebuilding different atmospheres as in Constant Nieuwenhuys's *New Babylon*. The archaic logic of ground, floor, walls, roof, and sky gave way to that of the mobile "package." Buildings became small bubbles of controlled space. The kit could be as minimal as a high-tech set of clothes, like Michael Webb's

Suitaloon or David Greene's *Inflatable Suit-Home*. And it was soon reduced down to a headset, a miniature enclosure that wrapped the eyes and ears to provide a total architectural experience, as in Walter Pichler's *Portable Living Room* or Haus-Rucker-Co's *Mind Expander*. Even the headset was made redundant by the most extreme end game of this line of experiment, the *Architecture Pill* of Hans Hollein. A single dose is swallowed for the complete architectural experience. The building as a vehicle becomes architecture as a psychedelic trip, space-making as a branch of pharmacology. Architecture is whittled down to a tiny gel wrapper for some chemicals, a tiny space capsule.

This assault on the very idea of a building was launched simultaneously on many fronts. There were key units like Archigram in England, the Metabolists in Japan, Utopie in France, Haus-Rucker-Co and Coop-Himmelb(l)au in Austria, and Italian teams like Superstudio and Archizoom Associati. An international set of soloists also played a key role in the campaign: Paolo Soleri with his futuristic cities for outer and inner space; Frei Otto's netted cities and bubble worlds; Gunther Domenig's vast lattices serviced by blobby modules; Eckhard Shultze-Fielitz's sprawling triangulated *Space City*; Yona Friedman's suspended *Mobile City*; Francois Dallegret's *Un-House*, and so on. Most of the operatives were teachers, so a whole network of students joined in. The combined effect was that of a global think-tank, a dispersed laboratory for a relentless wave of experiments. Its operations were closely monitored by newly launched little magazines, special issues of professional journals, and books with titles like *Fantastic Architecture*, *Experimental Architecture*, *Urban Structures For the Future*, *Anthropods*, *Megastructure: Urban Futures of the Recent Past*, *Where Shall we be Living Tomorrow?* and *Architecture 2000*. Key exhibitions like the 1967 *Urban Fictions: Visionary Architecture* in Austria reinforced the sense that a new kind of architectural fiction had been developed, a science fiction of high-tech worlds in which completely different kinds of life could be lived.

All the projects aimed at some ideal future by extrapolating the potential evolution of experimental technologies of the present. This fantasized hyper-evolved architecture didn't simply transport humanity into new kinds of space. It produced new kinds of humans. The escape from

HAUS-RUCKER CO
HAVE A "PSY-YEAR," 1968

KEVIN ZUCKER
ANGELS; THE HEADS OF PINS, 2000
ACRYLIC, INK-JET AND CARBON TRANSFERS, AND ENAMEL ON CANVAS
60 X 45 INCHES
COLLECTION OF HUGH J. FREUND, NEW YORK
COURTESY OF MARY BOONE GALLERY, NEW YORK

the ground was also the escape from the limits of our fleshy bodies. Humanity was redesigned. Traditional city and building types had to be dissolved because they monumentalize the characteristics of a soon-to-be-extinct species trapped in obsolete social patterns. In the representations of new suspended worlds, traditional structures can sometimes be seen, falling into decay down below or off in the background, menacingly dark and static in comparison to the glistening see-through forms of mobility that have arrived. But usually the architecture of the present had simply vanished. To look at an architectural drawing was to look at the future through a very high-grade crystal ball. Architects eagerly invented a slew of labels for the radical worlds they exposed: "Biotecture," "Infotecture," "l'architecture mobile," "Parametric Architecture," "Nuclear Architecture," "Chemical Architecture," "Ville cybernetique," "Arcology," "Metabolist Architecture," "Interplanetary Architecture," and so on. The future was engineered in the global think-tank, circulated among friends, then thrown at a fascinated but largely skeptical architectural community.

A primary source for much of the imagery was the space race, which occupied so much of the mass media during those years. Architects took inspiration and an aura of credibility from the regular stream of NASA images of colossal assembly buildings, mobile launching pads, sleek capsules, wrinkled spacesuits, and the geometric perfection of revolving space stations. The spacesuit became the prototype of a new kind of house and the space station became the prototype of a new kind of city. Astronauts appeared all over architectural publications. Indeed, the point of much of the design work was to turn us all into astronauts, nomadic explorers of the outer limits whose bodies are symbiotically entangled with the high-tech mechanisms of our architecture. The reality of NASA's mission to produce an architecture released from gravity launched waves of speculative fantasy. In the hands of the architects, this emerging reality merged seamlessly with the perennial themes of science fiction.

Science fiction has always been a rich source of architectural images. Ever since the classic stories of Jules Verne and H.G. Wells, there has been a trajectory of detailed fantasies about cities and buildings in sci-fi narratives

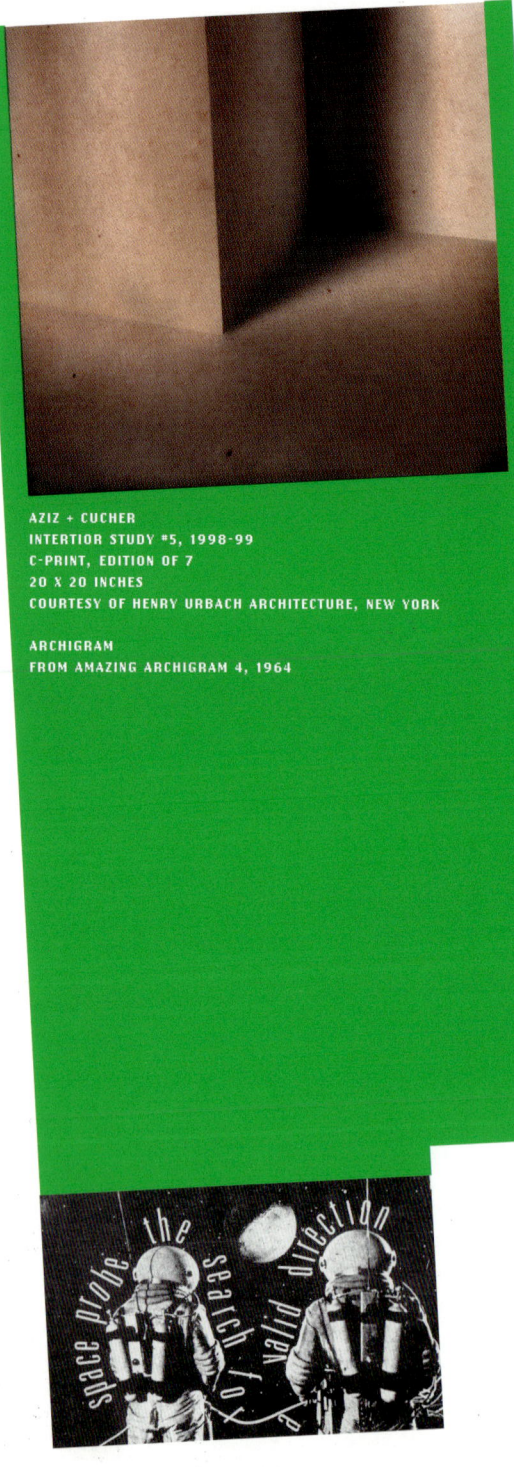

AZIZ + CUCHER
INTERTIOR STUDY #5, 1998-99
C-PRINT, EDITION OF 7
20 X 20 INCHES
COURTESY OF HENRY URBACH ARCHITECTURE, NEW YORK

ARCHIGRAM
FROM AMAZING ARCHIGRAM 4, 1964

that runs in parallel to the contemporary visions of architects. In the sixties, many designers simply tried to remove the gap between these two trajectories. The only difference between reading a science fiction story and an architect's project was that the architect might use a few words to clarify the images while the storyteller might use a few images to clarify the words. But in becoming science fiction writers, architects presented ever more elaborate narratives about their future worlds. Superstudio, for example, developed an entire genre in which an immaculate image is accompanied by an extremely detailed narrative about a radical form of space and life at the end of civilization as we know it. While traditional architecture disappears, strange forms of humanoid walk on the beach; cyborgs evolve in glistening grids; people commune with death by plugging into computers; a collective brain orbits earth; an ideal city revolves in deep space; and so on. Many of the Superstudio narratives were so detailed that they were presented as movie scripts with storyboards when first published in architectural magazines, and a number of these architectural sci-fi films were made. Sophisticated commentary about architecture was offered in the disguise of futuristic stories.

Many of the science fiction architects literally set their stories in outer space but those who buried their projects in the ground or suspended them just above it were equally space-age. Already in 1959, for example, Constant described the endless web of scaffolding covering the planet in his *New Babylon* as a kind of "architectural science fiction,"[1] and the seminal 1962 issue of *L'Architecture d'aujourd'hui* on "architectures fantastiques," likewise described Shultz-Fielitz's megastructural *Space City* as "realizable science fiction urbanism."[2] Even Michel Ragon's 1963 survey of the experimental scene describes itself on the back cover as being "as passionate as a science fiction novel."[3]

The most obvious example is the fourth issue of the little magazine *Archigram* from 1964 on "Zoom-Architecture," which is presented as a space comic collaged out of frames from different science fiction strips. Extra bubbles are added so that the new strip can reflect on its own role. One of the figures in a futuristic settlement now insists that the space comic universe can "inspire and encourage the emergence of more courageous concepts" of the city.[4] Another character looks out at the reader and announces that "the strip cartoon provides a visual jump-off point—a mental zoom boost—enables us to push aside architectural waste-matter so that reality may emerge."[5] Science fiction is seen to inspire new forms of invention by changing the sense of what might be real. The whole issue of the magazine is an attempt to blur the line between reality and fantasy, fact and fiction. Yet another character in the strip turns to us and says that "close examination of SPACE COMIC material reveals a two-way exchange between space comic imagery and the more advanced 'real' concepts and prophecies—Geodesic nets, pneumatic tubes, plastic domes and bubbles—the world of the thinks-balloon and the inventor's pad overlap."[6]

When the strip ends, designs by Archigram members appear in the middle of images of rockets, astronauts, lunar landscapes, and deep space. Superman turns into an architectural critic, flying over Peter Cook's *Plug-In City* project and offering his informed opinion. There is even a pop-up centerfold featuring some of the group's tower designs that rise up against the background of the futuristic city from the strip, and a few blasting rocket ships. Archigram projects get associated with a wide range of contemporary experimentation and are suspended between images labeled "science fiction" or "science fact." The little magazine systematically challenges the divisions between "so called 'real architecture,'" projects for the "near future" and sci-fi architecture of the distant future. It calls for a radical "cross fertilization" between architecture and science fiction and even ends by giving a list of recommended science fiction readings for architects.

This blurring has a precise history. It started the year before in Archigram's lesser-known *Living City* installation at the ICA gallery in London, a city simulator whose dense all-enveloping three-dimensional collage incorporated science fiction figures like Superman, the superwoman robot from Fritz Lang's *Metropolis*, and Robby the Robot from the movie *The Forbidden Planet*. In the accompanying publication, a space comic character has an idea near a do-it-yourself medieval cathedral and a NASA astronaut celebrates being creatively nonproductive. The voices of the group members David

Greene and Michael Webb are shown coming out of the mouths of astronauts alongside "The Story of the Thing," their narrative of a "science fiction city," a vast net coming from outer space and encircling the earth for an age in which people no longer need to live in cities because of the new forms of electronic communication.[7]

One of the acknowledged role models for the *Living City* show was the famous installation featuring Robby the Robot by some of the so called Independent Group in the *This is Tomorrow* exhibition of 1956. Almost all the Independent Group described themselves as sci-fi fans, avid collectors of cultish sci-fi magazines and viewers of B-grade sci-fi films. Eduardo Paolozzi was incorporating covers of science fiction stories in his scrapbooks of collages around 1952. Lawrence Alloway formally lectured to the group on science fiction in January 1954, proposed an exhibition about it in 1955, and published articles on it from 1956 on. John McHale frequently referred to science fiction, published images from his favorites, and wrote introductions to collections of sci-fi stories. Like his fellow Independent Group members, Magda Cordell and Paolozzi, he even produced a kind of science fiction imagery. They bred whole new species of mutants and androids in their paintings, sculptures, and collages while obsessing about robots, monsters, and cyborgs in their writings. Reyner Banham was likewise hooked, frequently using it as a reference point, and even publishing a 1958 piece calling on architecture students to literally produce science fiction architecture.[8]

Archigram, whose base was literally over the road from Banham's house, seemed to take up the call. Yet their *Living City* publication notes how the space program had fundamentally changed the status of the Robby the Robot figure so famously incorporated into *This is Tomorrow*. They differentiated themselves from the Independent Group by concentrating on the functional organizations offered in science fiction imagery rather than analyzing their iconographic resonances. The point was to try to realize the images rather than explore their fantasy role in contemporary society, since NASA had both given a new sense of reality to once fantastic images and created a demand for new kinds of images. McHale had increasingly moved in a similar direction, as exemplified by his "2000+" special issue for *Architectural Design* in 1967. It presented many images from the space program, interspersed with classic science fiction graphics from the seventeenth century on, when trying to predict the future character of settlements and life styles—with particular attention to the central role of the computer.[9]

In this, McHale's thinking moved in tandem with the maverick guru Buckminster Fuller, whom he had been writing about since the mid-fifties, becoming his first biographer in 1962 and actively collaborating with him on research projects between 1962 and 1968. In fact, the whole science fiction tradition of the sixties is indebted to Fuller. His structural systems, domes, and geometric patterns either literally made their way into much of the work or his experiments were cited by architects as an inspiration. But more than this, he was a role model in creatively negotiating between fact and fiction. His very first book, *Nine Chains to the Moon* of 1938, ends with a science fiction story—"Jones and the X-ian"—in which a man developing an ultra short-wave receiver makes contact with a woman from planet 80 XK 23 in trapezoidal segment 727831 of the star layer.[10] She communicates directly with his brain and reveals the philosophy of her spectacularly advanced world and the different attitude that philosophy has to "scientific dwelling machines," the theme of the whole book. Fuller's own position effectively gets credited to an infinitely superior alien from the far side of the universe. All of his theories and design work can be understood as a form of science fiction, driven by a constant sensitivity to communication systems and an ultimate belief that the real architecture of the cosmos is invisible.

It is Fuller, after all, who redefined the planet as "Space Ship Earth," turning us all into astronauts moving at great speed through space and making every architectural project, no matter how earthbound, a part of spaceship design. Fuller was if anything most at home in the science fiction world—a number of its best known writers became close friends. Despite his cult status in architecture, very few designers wanted to go as far as he in visualizing buildings as merely a primitive communication system that will have to give way to an invisible architecture for a species that has yet to discover that it can communicate with pure brainwaves. As he put it in 1970: "All good science fiction develops real-

NINA BOVASSO
UNTITLED, 2000
ACRYLIC ON PAPER
27 X 39 INCHES
COURTESY OF PAUL KASMIN GALLERY, NEW YORK

SUPERSTUDIO
INTERPLANETARY ARCHITECTURE, 1972

CRAIG KALPAKJIAN
ROOM, 1996
CIBACHROME PRINT ON ALUMINUM
29 1/2 X 39 1/2 INCHES
COURTESY OF ANDREA ROSEN GALLERY, NEW YORK

HALUK AKAKÇE
STILL LIFE, 2002
INSTALLATION WITH TWO-CHANNEL VIDEO PROJECTION
DIMENSIONS VARIABLE
COURTESY OF BERNIER/ELIADES, ATHENS, AND DEITCH PROJECTS, NEW YORK

istically that which scientific data suggests to be imminent. It is good science fiction to suppose that a superb telepathic communication system is inter-linking those young citizens of world-around."[11] While waiting for the development of invisible architecture, Fuller led the campaign to lighten architecture. From his first designs of the late twenties for "lightful" houses that would be carried from site to site by Zeppelin, to his porous geodesic domes carried by helicopter, to his proposals for vast spherical transparent balloons circumnavigating the globe and carrying whole neighborhoods from mountain top to mountain top, he tried to dematerialize buildings as much as possible. It was inevitable that he would become the patron saint of the sixties science fiction wave. It is symptomatic, for example, that shortly after referring to Fuller's project at a 1960 lecture for covering Manhattan with a transparent dome, that Constant calls for a "technological science fiction"[12]

The sense of the sixties architect as essentially a fiction writer never faded. It just suddenly stopped. Almost nothing of all the countless dreams of what was supposed to be our near future was ever built. And in the early seventies the whole science fiction genre of architecture was abruptly discarded, only to be equally abruptly revived in the nineties when the latest generation started to echo much of the same mentality. The drive towards dematerialization was relaunched and the future of architecture was again seen in the computer. Even some of the formal vocabulary was echoed. Yet the new architects acted as if the sixties experiments never took place, and didn't see their own work as science fiction. If the reality of the space program had transformed the status of classic sci-fi fantasies, the reality of the computer in today's society has transformed the status of all the sixties fantasies about how the computer would revolutionize architectural experience. Contemporary architects abandon playful fictions in favor of pseudoscientific discussion of the algorithms used to generate their latest shapes. Yet it is precisely there, in the earnest attempt to give projects the aura of scientific experimentation, that the science fiction genre has been unconsciously revived. It is precisely when architects appear in lab coats rather than artist gear that the fiction is extended. Narratives are skillfully developed that give dreams about an idealized distant future the sense of not just being in the very

immediate future, but the sense of having already happened. After all, as Orson Welles's infamous radio broadcast of 1938 demonstrated, science fiction takes its strongest form when disguised as documentary.

It is not simply that the last decade has seen an unconscious revival of the science fiction genre of the sixties. The sixties was itself such a revival. Indeed, there is a whole genre of such projects, a tradition of architects visualizing worlds of the future, that is usually labeled "visionary," "utopian," "ideal," "futurist," or "fantastic." This legacy was clearly understood by Archigram, who saw themselves resurrecting an attitude from the first decades of the century. In the *ZOOM* issue, expressionist visions from 1919 to 1921 by Bruno Taut, Carl Krayl, and Hans Luckhardt are compared to the images of Archigram's own work, the space program, and sci-fi comics—"proving" that architects "can be as wild, and as dynamic as the cartoonists." A text by Peter Cook highlights the resonance between the latest architectural experiments, science fiction graphics, and the dreams of the early twentieth-century architectural avant-gardes—while pointing out that architects usually lag behind the sci-fi writers.[13] The group's own mission is to narrow the gap by occupying the "never land between gesture and architectural laboratory work," rejecting a clear cut opposition of fact and fiction in favor of a continuum between the two.

The effect of undermining the distinction between fact and fiction is both to give the latest fantasies a sense of realism and to facilitate the development of even more radical fantasies. And pointing out the precedents has the advantage of making the current moves part of an ongoing research program. Not by chance were most of the sixties experiments in futurism preceded by the retrospective surveys in the *Visionary Architecture* show curated by Arthur Drexler at MoMA in 1960 and the well-known book *Phantastische Architektur* by Ulrich Conrads and Hans Sperlich of the same year.[14] Looking back at previous experiments documented in those surveys (or other projects that were visualized as orbiting planets, like Wenzel Hablik's *Colony-In-the-Air*, a flying city of 1908, and Malevich's flying *Suprematist Architon* structures of the twenties) can set the stage for variations of old science fiction ambitions. Greene and Webb's

MATTHEW NORTHRIDGE
NEW CITY (DETAIL), 1998-2002
PAPER AND MASONITE, CONSISTING OF 3000+
INDIVIDUAL PIECES
24 X 96 X 96 INCHES
COURTESY OF THE ARTIST
PHOTO: TOM VAN EYNDE

1963 "Story of a Thing," for example, begins with an image of Friedrich Kiesler's *City in Space* from thirty-eight years earlier and a citation from Kiesler's original description that was easily applied to their own vast net: "A system of tension in free space. A change of space into urbanism. No foundations. No walls. Detachment from the earth."[15]

Yet these ambitions go back much earlier than the twentieth century. To envisage architecture floating in space is an ancient dream. Indeed, it can be argued that the science fiction tradition is as old as the first architectural publications. Architecture, as distinct from "mere" building carrying out practical functions, is traditionally constructed as the representative of an ideal world. The logic of classical architecture is that a building is not simply a beautiful object in the world. It is an object whose precise proportions capture the transcendental harmonies of the cosmos so that it is at once in the world and beyond it, in time and beyond time, in culture and beyond culture. Architecture is understood to be a special bridge between the real and ideal. Architects fantasize about idealized forms then negotiate with real world constraints. Theoretical treatises are filled with perfect site-less objects and instructions on how to minimize their compromise when applied to particular sites. In reverse, architects usually remove evidence of compromise when they publish their work. Most projects come to life floating in the ideal abstract space of blank paper and end up there. Architecture keeps acting as an ambassador from a dream world.

This metaphysical ambition did not go away whenever classical architecture was rejected. The idealism of architects does not disappear when they switch allegiances from ideal geometries to the materialism of social life, technological evolution, vernacular tradition, economic forces, or popular culture. The idealism simply gets displaced. A kind of perfection is found in the supposedly imperfect and transitory world. New mythical images are assembled from even the most prosaic material. In the early decades of the twentieth century, for example, many radical dream worlds were concocted, futuristic scenarios made seductive by new kinds of stories and renderings, elements of which eventually made their way into the everyday environment and now seem utterly prosaic.

Some of the most innovative proposals have become easily overlooked default settings. There are standard labels for this kind of speculative project, like "city of the future," "city of tomorrow," "metropolis of tomorrow," and "house of the future." And there are standard sites for displaying them: world's fairs, idea competitions, exhibitions, and magazines—abstract spaces that relieve architects from the constraints of a particular material site. An active futurism permeates architectural polemics with designers routinely offering glimpses of tomorrow.

Yet this kind of fiction is often disguised as realism. Take, for example, the great storyteller Le Corbusier, arguably the architect who had the single biggest impact on the physical environment in the twentieth century. When he first presented his vision for the reorganization of the city in a 1922 exhibition, it was so radical that journalists kept describing it as "The City of the Future." He reports that his friends likewise said "All this for the year 2000!"[16] and "You work for the moon!"[17] but he insisted on calling it "A Contemporary City" and his presentation took the form of a hundred square meter photo-realistic diorama. As with most architect's projects, the proposal was made to look like an existing fact rather than an idealistic fiction. In a magic illusion, the scheme appears to have been found rather than designed. When publishing the theory behind it in the 1925 book *Urbanisme*, a long story that begins with a critical analysis of the way donkeys walk, Le Corbusier argues that he actually proceeded "in the manner of the investigator in his laboratory," assuming an "ideal site" to discover the general principles of an "ideal city" before applying them to the compromising conditions of any particular site.[18] He repeatedly refers to his ideas as "certainties" and insists that "this is no dangerous futurism, a sort of literary dynamite flung violently at the spectator."[19] But a disguised futurism it surely is, and he will eventually allow the book to be translated as *The City of To-Morrow* in 1929 when its images were already beginning to have their extraordinary effect on cities all over the planet.

Furthermore, this congenital futurism of the architect is always a kind of science fiction inasmuch as architecture is precisely that art whose unique identity partly derives from its supposed bond with science. Nowhere is the state-of-the-art of technology more discussed than in architectural discourse, despite the fact that buildings are actually extremely low-technology objects. Most household appliances are considerably more advanced than the houses that contain them. Buildings are usually the last thing to be updated by new technologies, yet architects are among the first to discuss each new technology. It is hard to think of a major scientific discovery that is not promptly cited by some architect or other as inspiration for their latest work. There is a huge gap between the speculations of the architect and the current realities of building, a science fiction gap. Architects become skilled at exaggerating the real into the fantastic. Leonidov's projects of 1928 and 1929 can imagine "radio-pictures," a vision of what we would now call television, with images transmitted around the world and projected onto huge outdoor screens; Frank Lloyd Wright can imagine a mile-high skyscraper in 1956; and William Katavolos can imagine self-molding plastic environments in 1960 that allow furniture, buildings, and cities to fluidly arise out of a kind of plastic foam and keep changing shape during the day to perfectly accommodate their occupants.

Yet these more extreme speculations are periodically expunged from the record for strategic reasons. All the constructivist, expressionist, and futurist experiments, for example, were completely suppressed when the official mythology of modern architecture started to be assembled in the late twenties—despite the fact that the radical experiments had played a key inspirational role, and many of the architects of the more practical proposals were the very same people who had carried out the now unmentionable experiments. In the end, all that remained of the intense fantasies about mobility and levitation was lifting buildings up off the ground with thin legs, and all that remained of the dream of dematerialization was the use of a lot of glass to produce transparency. Extreme imagery was displaced by seemingly practical low-tech proposals. A sustained burst of publicity gave way to instant amnesia. And the memories didn't return until the end of the fifties when mainstream modern architecture was constantly under attack and a new generation of fantasies was called for. Yet the same fate awaited the sixties science fictions. They were themselves soon forgotten and many of the architects responsible for them started to make considerably less provocative buildings. Perhaps one day we will say the same thing of the nineties digital fictions, whose arrival was symptomatically accompanied by a recollection of the sixties in retrospective exhibitions and publications.

An effect of this rhythmic oscillation between fantasy and practical reality is an imagined spectrum stretching continuously from the unbuildable futurism of "paper architects" through to the realism of architects who simply try to satisfy their clients by using the accepted techniques. Each architect can choose, at each moment, how far away to appear from the assumed practical, social, and aesthetic conventions of the day. Some are permanently associated with one end of the spectrum or another, but most move fluidly up and down the scale.

In the end, even the full-time futurists cannot simply be separated from the full-time realists. Architects are not really builders. They are storytellers, fiction writers. They tell wonderful stories about objects. They make us believe that huge assemblages of concrete, metal, wood, and glass can to come to life and say something about us and the world. The discipline is always more about fantasies than facts. The architect tries to realize fantasies but not just in the sense of turning them into solid material. Rather, it is a matter of making the fantasy itself seem real—as if it already exists. And if the project is actually built, the fantasy element remains. Indeed, precisely because it involves massive lumps of material, the sense of fantasy is accentuated. Even the most practical work by an architect involves a degree of speculation. A project is always placed beyond the given situation in a dream world, a space of desire, no matter how small the speculative leap. There is therefore a sense of science fiction in even the most humble project. Each idea is projected, thrown into an unknown future. The architect's stories inevitably aim ahead, while the buildings almost always lag behind contemporary technical and conceptual developments. Architectural discourse is ultimately a particular way of negotiating between past and future. It is a way of looking forward and backward at the same time.

In this sense, the latest digital experiments simply take to an extreme the architect's normal role as creative science fiction writer. Whether the flying jellyfish remain inside the computer, or crash-land onto particular sites, or their architects move on to safer proposals, is not the issue. Cyborgs, intelligent membranes, and topological geometries returned to invade much of the recent discourse but they are not so futuristic in the end. On the contrary, they mark a refreshing return to crucial but abandoned lines of experiment. Some of the mentality and aesthetic of an earlier mode of science fiction has simply been revived. And even the most innovative, radical, or impractical ambitions of the latest experiments are finally no more futuristic than when a seemingly down-to-earth architect turns to her or his client with a gleam in the eye, pointing to a drawing, model, or building and speaking lyrically about how a particular detail creates a sense of openness, tension, responsiveness, modernity, or whatever. The real science fiction is to be found in the everyday miracles of the architect as an expert storyteller conjuring up magic properties—making fantasies seem real and animating inert objects.

1. Constant, "Le grand jeu à venir," *Potlach. Informations intérieurs de l'Internationale Situationniste*, no. 1 (no. 30): n.p.

2. Eckhard Shulze-Fielitz, "Une théorie pour l'occupation de l'espace," *L'Architecture d'aujourd'hui*, no. 102 (June–July 1962): 84.

3. Michel Ragon, *Ou vivrons-nous demain?* (Paris: Robert Laffont, 1963), back cover.

4. Archigram, *Amazing Archigram* 4 (1964): 1.

5. Ibid., 4.

6. Ibid.

7. David Greene and Michael Webb, "The Story of the Thing," *Living Arts* no. 2 (1963): 92–93.

8. See, for example: Lawrence Alloway, "Technology and Sex in Science Fiction," *Ark*, no.17 (Summer 1956): 19–23; John McHale, "The Expendable Ikon 1," *Architectural Design*, vol. XXIX, (February 1959): 82–83; Reyner Banham, "Space, Fiction and Architecture," *The Architects' Journal*, no. 3294 Vol. 127 (April 17, 1958): 559–560.

9. John McHale (ed.), "2000+" special issue of *Architectural Design*, vol. XXXVII (February 1967): 64–93.

10. Buckminster Fuller, *Nine Chains to the Moon* (New York: Lippincot, 1938).

11. Buckminster Fuller, "Introduction," in Gene Youngblood, *Expanded Cinema* (Toronto: Clarke, Irwin and Company, 1970), 15.

12. Constant, "Was ist Städtebau?" manuscript of lecture given at Institut für Städtebau und Landesplanung of the Rheinische-Westfälische Technische Hochscule, Aachen, July 18, 1960.

13. Peter Cook, "Zoom and Real Architecture," *Amazing Archigram 4* (1964): 18.

14. Ulrich Conrads and Hans Sperlich, *Phantastische Architektur* (Stuttgart: G. Hatge, 1960).

15. Friedrich Kiesler, "City in Space" (1925) cited by David Greene and Michael Webb, "Story of the Thing": 92.

16. Le Corbusier, *Urbanisme* (Paris: Crès, 1925), iii.

17. Le Corbusier, *Oeuvre Complète*, vol. 1. (Erlenhach: Boesiger, 1927), 34.

18. Le Corbusier, *Urbanisme*, 158.

19. Ibid., 168.

PATRICK MEAGHER
DIGITAL PRINTS FROM UNITBEAD 2.0
USER-NAVIGABLE DATA PROJECTION INSTALLATION:
G4 ATRIUM, 2001
TEKTRONIX PARK, 2001
HP WAY, 2001
EPSON PRINTS
16 X 20 INCHES EACH
COURTESY OF THE ARTIST

Although the roots of virtual reality reach as far back as the 1960s, it did not become a popular topic of public conversation until this past decade. Beginning with cinematographer Morton Heilig's *Sensorama Simulator* in the 1960s, developments in virtual reality were driven by the U.S. military's experimentation with flight simulators in the 1970s, the entertainment industry's creation of video games in the 1970s, computer-generated movies in the 1980s, and the expansion of cyberspace in the 1990s. Now, virtual reality labs are popping up at major universities all over the world and aggressively initiating significant studies in the fields of medicine, education, communication, engineering, and architecture. True virtual reality is still in its infancy and few individuals have experienced full-body immersion, yet many can imagine and hypothesize about its effects based on their own experiences of movie theaters, video games, and the computer.

Patrick Meagher's computer-navigable three-dimensional animation of a futuristic city-scape is about as close as an *Out of Site* visitor is going to get to a fully immersive virtual reality experience without donning special headgear. Most often, virtual reality is associated with projection goggles, wired body suits, and sensor gloves that pass on movements to a computer, triggering the graphics that give participants the illusion of being submerged within a computer model. A telepresence, however, can be generated by the simple use of a computer and mouse, to create the experience of being fully present in another world remote from one's stimuli. Meagher's standard screen-based interface presents three life-size projections of architectural spaces that fuse with the gallery's physical structure to form a panoramic spatial simulation. Though this manufactured reality falls short of our expectations of high-degree realism, Meagher's successful exploitation of shape, surface texture, lighting, perspective, depth of field, and anti-aliasing (the smooth gradation of pixels) associated with the creation of moving three-dimensional graphics, coaxes exhibition navigators into losing themselves in an unreal reality. What appears to be a seemingly superslick animated interior office, exterior building façade, or maze of corridors is in actuality a micro environment turned macro through the use of manipulated downloaded images of Styrofoam packing materials.

STEPHEN HENDEE
SILENT SECTOR, 2002
DIGITAL RENDERING OF SITE-SPECIFIC INSTALLATION OF COR-X, TAPE, AND LIGHTING
DIMENSIONS VARIABLE
COURTESY OF THE ARTIST AND HENRY URBACH ARCHITECTURE, NEW YORK

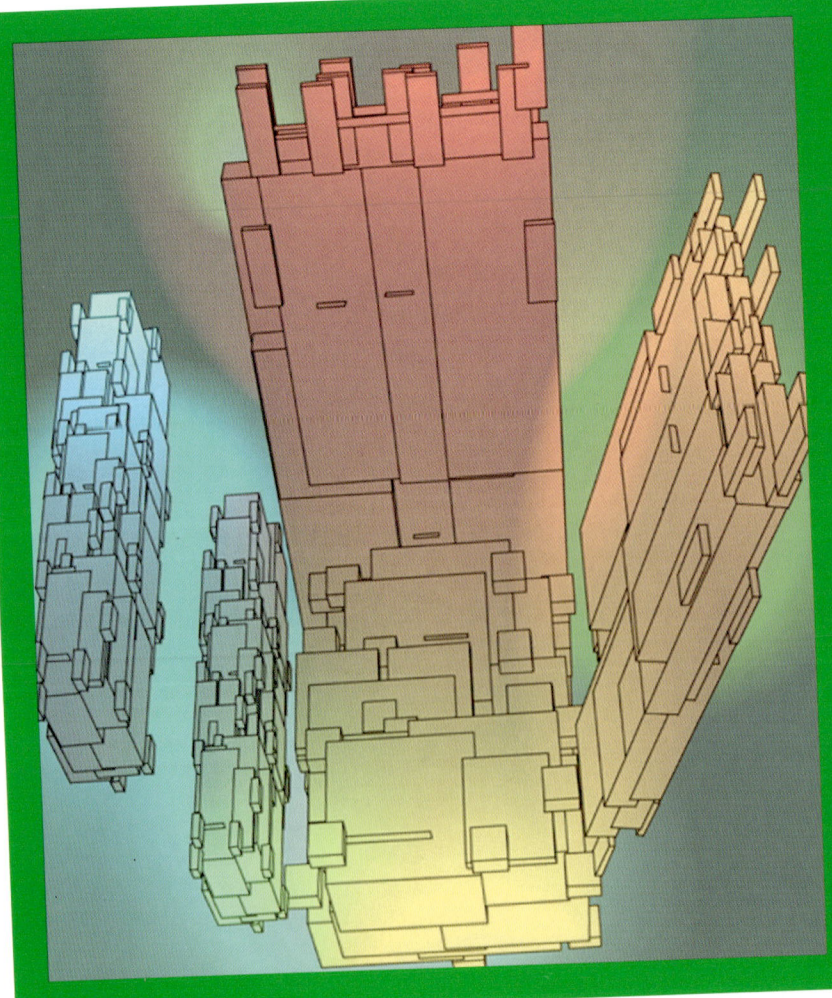

The concept of virtual reality typically invokes high-tech visions of computer animations, but many broad artistic and literary interpretations proliferate. Virtuality can happen when you are reading a book, sitting in a movie theater, talking on the phone, riding an amusement park ride, playing a video game, or surfing the Internet. It is, in fact, any activity that transports you into a world of information or supplies an illusion that conceptually catapults you into distant realms and spaces. Cyberspace, the term that originated in William Gibson's 1984 sci-fi novel, *Neuromancer*, literally means "navigable" space. The cyberspace of Gibson's videogame-inspired fictional matrix of quasi-dimensions is no longer science fiction; it is an extension of the idea of virtual reality and accepted as a part of our everyday reality. Today, cyberspace refers to the imperceptible spatial links on the Internet and is applied to most virtual experiences created in a computer. It is less about converting data into pictures that come from human experience, i.e. in flight simulation, and more about the abstraction of hardware, software, and data. Numerous historians and theorists have recently referred to cyberspace as the new frontier—a vast, unmapped, culturally and legally ambiguous land ready for exploration—much like the unsettled nineteenth-century West.

Just as the railroads opened eyes and minds to the uncharted American West, causing a sudden change in perceptions of time and space, cyberspace, the Internet, and virtual reality have reshaped our understanding of the world by posing tough questions about philosophies of space, challenging common spatial perceptions, and reexamining accepted definitions of architecture. As Martin Heidegger warned, new modes of media have created the collapse of traditional notions of time and space. Absolute space, frequently described as a container filled with objects, differs from the space of cyberspace. Cyberspace is relational and is the consequence of interactions. It is about the flow of information within space rather than the old space of places.[1] The artists in *Out of Site* demonstrate the cybernetic impact on architecture by engaging in what Marcos Novak calls "eversion"—the outpouring of virtuality onto ordinary space[2]—to create virtually real spaces of impermanent materials which we can inhabit physically and/or psychologically.

Novak presents eversion as an acceptable alternative, a complementary concept to the notion of immersion, which attempts to separate the user from the real world. To evert virtual reality is to turn it inside out. While eversion, like immersion, addresses our entry into information spaces, it also accounts for a casting outward of the virtual onto the space of everyday experience, which Novak argues, we are unlikely to abandon any time in the near future. Stephen Hendee's installations are quite visibly eversionistic. Not only does he take almost every identifiable aspect of virtual reality space and supplant it onto actual space, he literally turns it inside out. He spontaneously assembles foam core, black electrical tape, and fluorescent lights to form retro tunnels and caves that resemble the innards of an outdated glowing phosphor green computer monitor. The spaces recall the domed city from the low-tech 1976 movie *Logan's Run*, where occupants were sealed inside the city by computers attempting to maintain a population balance, or *Tron* (1982)—one of the first computer-generated/videogame-inspired movies where a hacker is transported inside a computer. Similar to the spaces presented in *Logan's Run* and *Tron*, Hendee's anonymous, futuristic, architectonic environments are complex webs of exponentially mutated cells that reveal numerous vectors, multiple dimensions, and skewed perspectives. True to Novak's eversion phenomenon, Hendee's space is a compilation of local, distant, and virtual spaces.

Is Hendee's space real or fictional? The "real" is an idea that we are constantly reevaluating and virtual reality is actively pushing the edges of what we currently know to be reality. The French philosopher Paul Virilio argues that the "real" and the "virtual" are distinct, but that new technologies are beginning to substitute a virtual reality for an actual reality. Sociologist Jean Baudrillard claims that virtual reality is as real as anything we normally call reality and that the virtual space will present an increasingly more real space. Both theories end up in the same place, proposing that the virtual or the hyperreal will replace the kinetic real space that it once simulated.[3] Neither hypothesis proves to be adequate when applied to the virtually real eversionist architectures of the *Out of Site* artists. According to Gilles Deleuze, we confuse issues when we compare the "real" with the "virtual." We should instead compare the "virtual" with the "actual" and the "real" with the "possible."[4] Thus an event can be both real and virtual. Deleuze states:

Indeed, the virtual must be defined as strictly a part of the real object—as though the object had one part of itself in the virtual into which it plunged as though into an objective dimension.[5]

Pierre Lévy too believes that the virtual is by no means the opposite of the real:

...the virtual, strictly defined, has little relationship to that which is false, illusory or imaginary. On the contrary, it is a fecund and powerful mode of being that expands the process of creation, opens up the future, injects a core of meaning beneath and the platitude of immediate physical presence.[6]

Whereas the twisted, multiplied, and slightly warped elements of cyberspace—results of isometric perspective, poor resolution, device latencies, and other shortcomings—resemble the visuals we associate with the fictional, for millions of people cyberspace is a real place. Now in the twenty-first century, we have successfully integrated various realms of virtual reality, such as television, film, and the Internet, into our lives. Cyberspace is no longer a non-space or a paraspace (separate space)—the virtual intersects with the real or more precisely the virtual has been assimilated into reality and reality is changing.

One example of how reality is changing is with the notion of viewer-centered perspective, which is perhaps one of the most interesting phenomena implicated by cyberspace. For the first time since the Renaissance, perspective has been redefined. Many of us, on a daily basis, sit in front of our computers, manipulating their contents with a mouse and screen—simultaneously opening multiple windows that project widely varying bits of information. We can circumnavigate a building from our desk chair in a matter of seconds. We look in one direction, but are able to view multiplied perspectives and slip between liminal dimensions. The effect of this is particularly tangible in Seattle artist Victoria Haven's large sculptural drawings constructed out of various combinations of scotch tape, acetate, pins, paper, ink, paint, shadow, light, and reflection (shown only in the Henry Art

Gallery's installation of *Out of Site*). Haven's deceptively simple installations appear, at first glance, to be straightforward, slightly overblown, preparatory architectural studies. They quickly morph, however, into self-sustaining structures that live somewhere between the second and third dimensions. After further inspection, one realizes that Haven's buildings, like those in cyberspace, have more dimensions than exist in our physical environment. Similar to Hendee's installations, Haven's networks of odd geometric modular pieces are stuck together with an unconventional sense of order, building off each other to form obsessively disjunctured, moving spaces careening off into the distance. Sadie Plant, while writing about *The Meshed City*, commented:

The computer has multiplied and magnified the perceptible world, rendering much that was imperceptible perceptible, and exposing molecular complexity inside previously large-scale, simple things. It has revealed repeating patterns, processes extending across complex systems of many different scales and kinds.[7]

Haven's proliferating network of cells resembles an anonymous futuristic mobile city white-washed by speed and technology. These multiperspective and cross- or multidimensional spaces are not simply breaking the rules of perspectival format, they are part of a contemporary system reflecting new laws of existence.

"Liquid architecture," another concept discussed by Novak, falls into the "new laws of existence" category as an architecture influenced by cyberspace. Novak believes:

The ascendancy of the liquid over the fixed is a phenomenon that is emblematic of an intellectual condition in which previously irreconcilable, even inconceivable, opposites coalesce into strange but ultimately tenable alloys.[8]

Cyberspace, as a form of unsolid electronic space, leads us to rethink the stability and permanence of architecture. More and more architects are beginning to address the notion of continual motion and account for an environment that may quickly change its position or attributes, as in the architectures found in the movie the *The Matrix*. These new liquid architectures encourage warpage, tunneling, and distortion, altering the viewer's sense of physical space.

SVEN PÅHLSSON
WITH ERIK WOLLO
SPRAWVILLE OR LIFE AT HIGHWAY EXIT RAMP, 2001
3D ANIMATION WITH SOUND
10 MINUTES
COURTESY OF SPENCER BROWNSTONE GALLERY, NEW YORK

VICTORIA HAVEN
SUPERMODELCITY, 2000
TAPE, VELLUM, AND PINS
90 X 138 X 1 INCHES
COURTESY OF THE ARTIST
PHOTO: SPIKE MAFFORD

Ricci Albenda's white plastered Fiberglas portals blend into corners like chameleons. They function as elements of liquid architecture that bulge, inhale, and undulate—not with the click of a mouse or with the donning of headgear, but with the bat of an eye or the twist of a torso. Albenda doggedly distorts the viewer's perceptions—what at first appears to be an ordinary white cube is really not a white cube, but a cube of a cube, or a nod to another dimension. In Albenda's world, the space of flow begins to replace the space of place.

The space of flow is not simply about the potential liquidity of architecture or architecture in motion, but about the flow of data, upon which the world is constructed. Information technology is radically changing society—the physical no longer takes precedence over how data is exchanged. Stephen Hendee in a recent interview stated, "I think that contemporary architecture has much more to do with data than with permanence. Currently data is the most important thing in the world and it is one of the most flexible resources."[9] Albenda too is interested in information; more specifically, he is curious about our relationship to words, asserting that words have instant power to transport one to another place. His singular, optically skewed floating words applied directly to the wall are virtual objects. As with ancient Mesopotamian text, his words ask viewers to construct a space of meaning, supporting the idea that reality is based on symbols and language, not on reality as "it" is. Albenda is interested in exploring the possibility of virtually engaging ideas, abstractions, and words. He makes the invisible visible by subtly adapting spatial and/or written information and bending it to meet reality.

Albenda's physical installation of words resemble the information landscapes available in hypermedia browsers or found in the geographies of cyberspace. Cyberspace is not laid out the way we are used to experiencing our environment. It is an infinitely mutable form of information space that changes daily, even hourly, thereby constantly creating new geographies—maps begin to lose their accuracy as soon as they are printed. This was intentional. The complex web of information located on the Internet was originally developed by military communications in a decentralized fashion so as to prevent a central point of attack. As a result, it functions as a tracing map of connectivity that lets the casual user actively navigate various paths and gather information from multiple points of entry. Several entities merge, each traveling along its own route, and the Internet becomes a giant spatial fragmentation that is a layered, nonlinear, unstructured, and uncentered space that seems to run in every direction—the exact opposite of our current cities. As the theorist Sadie Plant acknowledged to Zoey Kroll:

The Internet is a place where you can see the possibilities of an ordered disorder on a self-organizing system. It obviously is not just chaotic, but it doesn't have a structure in a traditional sense, and that gives a lot of opportunity for rethinking, for example, the idea of the city.[10]

Julie Mehretu's explosive layers of complex, fragmented spaces recall Internet connectivity maps that echo the complex system of the Internet and its navigation. Her Mylar, vellum, paper, ink, paint, and acrylic-veiled images of socially charged spaces such as government buildings, airports, schools, and stadiums, are sprawling, stacked, and layered in such a way that the two-dimensional screens become moving three-dimensional spaces. Mehretu's work, like the Internet, is transformative and addresses a changing society; it visually parallels the restructuring of social, cultural, political, institutional, and economic life brought on by the Internet.[11]

Spatiality is at the core of human existence, allowing us to make sense of the world. G. Foy writes, "It does not matter if your space is psychological or physical: it is good for the ass to know where it is in space."[12] Architecture, in its reference to the physical world, offers clues to making the virtual world easier to understand and navigate. The reverse may also be true. If architecture is built on information and if spaces are defined as an array of information, then we might surmise that virtual reality, as a net of information, is drastically influencing our definition and understanding of physical architecture. According to Novak, "The architecture of cyberspace offers the opportunity to mend the rupture between our knowledge of the world and how we conceive and execute architecture. It allows a far greater latitude of experimentation than any previous architectonic opportunity."[13]

RICCI ALBENDA
PORTALS TO ANOTHER DIMENSION (MERSH)/POSITIVE, 2000
FIBERGLASS, EDITION OF 3
40 1/2 X 40 X 5 1/2 INCHES
COURTESY OF ANDREW KREPS GALLERY, NEW YORK

TOPOGRAPHICAL MAP OF THE INTERNET
EACH DEVICE ON THE INTERNET IS ASSIGNED A UNIQUE IP
ADDRESS RANGING FROM 0.0.0.0-255.255.255.255. FROM
A HOST COMPUTER IN NEW JERSEY, LUMETA SCANS THE ENTIRE
INTERNET, THE MACHINES THAT RESPONDED ARE DEPICTED HERE.
THE COLORING WAS DERIVED BY ASSIGNING EACH "OCTECT" A
PRIMARY COLOR. ROUGHLY 110,000 DEVICES ARE DEPICTED IN
THIS IMAGE, WHICH WAS HOW THE INTERNET LOOKED ON
JANUARY 24, 1999.
COURTESY OF LUMETA CORPORATION. PATENT(S) PENDING AND
COPYRIGHT LUMETA CORPORATION 2002. ALL RIGHTS RESERVED.

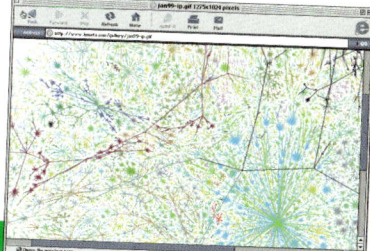

SHIRLEY TSE
BIONICPAK, 2001
EXTRUDED POLYSTYRENE
2 PARTS: 15 X 48 X 33 INCHES AND 31 X 48 X 33 INCHES
COURTESY OF MURRAY GUY GALLERY, NEW YORK, AND SHOSHANA WAYNE GALLERY, SANTA MONICA

THE "BILLBOARDS DISTRICT" IS A MULTI-AUTHORED VISUAL INFORMATION LANDSCAPE, FROM BITS AND SPACES (MARI ENGELI, EDITOR, BIRKHAUSER PUBLISHERS, BASEL, 2001)

Tangible spaces and their means and materials of construction are taking on new significance as a result of the growth of cyberspace. Many of the artists in *Out of Site*, while acknowledging the impact of virtual reality, have purposely chosen not to create the ultra-virtual. Instead they select simple, low-tech, commonplace materials, such as dry wall, tape, vellum, Styrofoam, and foam core, on which to transplant notions of virtual reality. The use of throwaway materials suggests adaptability and flexibility and hints at the impermanence of future architectures. Or perhaps the use of everyday materials represents the artists' assertion of the importance of physical materiality. For if, as some theorists suggest, the real will be usurped by the hyperreal, then there will be no need for physical structures. It's not an irrational premise—as it stands now, many of us do not have to leave our homes to take college courses, telecommute to work, talk with friends, order groceries, do our banking, or access myriad forms of entertainment. In 1831, a character in Victor Hugo's novel, *Notre Dame de Paris*, predicted that, "the book would kill the edifice." It didn't and it's unlikely that cyberspace will destroy physical architecture in the twenty-first century. While it is clear that reality as we know it is changing, we are not ready to give up the physical. The hybrid spaces created by these eversionist artists highlight materials that can be easily manipulated and molded to create skewed perspectives and new dimensions of layered spatial information. Uninhibited by the structural requirements imposed on traditional architects, artists are experimenting with the notion of liquid architecture and dreaming up new kinds of virtually real spaces. What was once considered imaginary or virtual is now becoming real. The virtual reality and cyberspace field is young—there are, as of yet, no rules and no concrete language to restrict one's imagination.

While it is thought which must explore the virtual down to the ground of its repetitions, it is imagination which must grasp the process of actualization from the point of view of these echoes or reprises. It is imagination which crosses domains, orders and levels, knocking down partitions coextensive with the world, guiding our bodies and inspiring our souls, grasping the unity of mind and nature; a larval consciousness which moves endlessly from science to dream and back again.[14]

1. Although used in a slightly different context, the "space of places" and the "space of flows" are ideas introduced by Manuel Castells in *The Information Age: Economy, Society and Culture. Volume 1: The Rise of the Network Society* (Cambridge, Mass. and Oxford, UK: Blackwell Publishers, 1996), 376–428.
2. Marcos Novak presents the term "eversion" in "Eversion: Brushing Against Avatars, Aliens and Angels," *Architectural Design* (September/October 1999): 73. This concept inspired the title and thesis of this essay.
3. Paul Virilio's and Jean Baudrillard's views on virtuality and real space are addressed in Jonathan Taylor's "The Emerging Geographies of Virtual Worlds," *The Geographical Review* (April 1997): 4, 7, and Mark Nunes's "Baudrillard in Cyberspace: Internet, Virtuality and Postmodernity," *Style* (i29m 1995): 314–316.
4. Gilles Deleuze, *Difference and Repetition*, translated by Paul Patton (New York: Columbia University Press, 1995), 208, 211.
5. Ibid., 208.
6. Pierre Lévy, *Becoming Virtual: Reality in the Digital Age*, translated by Roberto Bononno (New York and London: Plenum Trade, 1998), 16.
7. Sadie Plant, "The Meshed City," (internet essay). *Transarchitectures: Visions of Digital Communities*, 2.
8. Novak, 72.
9. Stephen Hendee, interview with Rochelle Steiner, *Wonderland* (Saint Louis, MO: The Saint Louis Art Museum, 2000), 69.
10. Sadie Plant, interview by Zoey Kroll, "Technically Speaking," (internet essay, June 1999): 3.
11. Martin Dodge and Rob Kitchin in "Introducing Cyberspace," *Mapping Cyberspace* (London and New York: Routledge, 2001), 13.
12. G. Foy, *Contraband* (New York: Bantham Books, 1997), 153.
13. Marcos Novak, "Transmitting Architecture: transTerraFirma/TidsvagNoll 2.0," *Architectural Digest* (November/December 1995): 43.
14. Deleuze, 220.

HALUK AKAKÇE studied art at Bilkent University in Ankara, Turkey, the Royal College of Art in London, and The School of the Art Institute in Chicago. He has had solo exhibitions at The Whitney Museum of American Art at Phillip Morris in New York (2002), Bernier/Eliades Gallery, Athens (2002), Deitch Projects in New York (2001), Centre d'Arte Contemporain, Geneva (2001), The Aldrich Museum of Contemporary Art, Ridgefield, CT (2001), Henry Urbach Architecture (2000), and P.S.1 Contemporary Art Center (2000) in New York. His group exhibitions include *Metropolitan Iconographies* at the XXV Bienal de Sao Paulo in Brazil (2002), *Animations* at P.S.1 Contemporary Art Center (2001), *Casino 2001* at the Stedejlik Museum voor Actuele Kunst in Ghent, Belgium (2001), *Painting at the Edge of the World* at the Walker Art Center, Minneapolis (2001), and the 6th International Istanbul Biennial, *The Passion & The Wave* (1999).

RICCI ALBENDA received his BFA from the Rhode Island School of Design in 1988. He has realized several exhibitions in New York and Europe, including solo exhibitions at Andrew Kreps Gallery in New York (2001), Van Laere Contemporary Art in Antwerp, Belgium (2001), and a project at The Museum of Modern Art (2001). Group exhibitions include *The Americans* at Barbican Art Gallery in London (2001), *Casino 2001* at the Stedejlik Museum voor Actuele Kunst (2001), *Greater New York* at P.S.1 Contemporary Art Center (2000), and *Glee: Painting Now* at The Aldrich Museum of Contemporary Art. He is a recipient of The Louis Comfort Tiffany Biennial Award (1999–2000).

AZIZ + CUCHER (Anthony Aziz and Sammy Cucher) have exhibited at the Herzlyia Museum in Israel (2002), Henry Urbach Architecture in New York (2001), the Museo Nacional Centro de Arte Reina Sofia in Madrid (2000), and at the Venezuelan Pavilion of the Venice Biennale (1995). They have participated in several group exhibitions including *Skin* at the Cooper-Hewitt National Design Museum, New York (2002), *Paradise Now* at the Tang Teaching Museum at Skidmore College in Saratoga, New York and Exit Art in New York (2001), *Sharing Exoticism* at the Bienniale de Lyon (2000), and *The Body in Art: 1950-2000* at the Arken Museum of Modern Art in Copenhagen (1999).

NINA BOVASSO received her BFA from San Francisco Art Institute in 1992 and her MFA from Bard College in 1999. She has had one-person exhibitions at Clementine Gallery, New York (2001, 1999, 1997), Jack Hanley Gallery in San Francisco (2002), Inman Gallery in Houston (2002), Richard Heller Gallery in Santa Monica (2001, 1999), the Fayerweather Gallery at the University of Virginia in Charlottesville (2001), and Vedanta Gallery in Chicago (2000). She has participated in group shows in Europe and the U.S. including *Terra Incognita: Contemporary Artists Maps & Other Visual Organizing Systems* at the Museum of Contemporary Art in St. Louis (2002), *The Sensational Line* at the Denver Art Museum, Colorado (2001), *Multiple Sensations* at the Yerba Buena Arts Center in San Francisco (2000), and *Monumental Drawings* at Exit Art (1999). In 2000, she received awards from the John Simon Guggenheim Memorial Foundation and the Marie Walsh Sharpe Studio Program.

STEPHEN HENDEE received his MFA from Stanford University in 1993. He has had solo exhibitions at Henry Urbach Architecture (2002), Mark Moore Gallery, Santa Monica (2001, 1999), Laguna Art Museum, California (2001), Rice University Art Gallery in Houston (2000), and the Southeastern Center for Contemporary Art in Winston-Salem, NC (2000). Hendee has participated in several group exhibitions including *Compression* at Feigen Contemporary, New York (2001), *Wonderland* at the Saint Louis Art Museum (2000), and *Generation Z* at P.S.1 Contemporary Art Center (1999). He will be in *Outer City, Inner Space* with Teresita Fernandez and Ester Partegas at the Whitney Philip Morris this summer. Hendee received the New Jersey State Council on the Arts award in 2000. He has also received The Louis Comfort Tiffany Foundation Grant, a Pollock Krasner Foundation Grant, a Marie Walsh Sharpe Foundation Studio Program award, and was a recent resident at the Headlands Center for the Arts.

CANNON HUDSON received his BFA from the California Institute of the Arts in 1988. He has exhibited at Galleria Marabini in Bologna, Italy (2000), Acme Gallery in Los Angeles (1995), Lauren Wittels (1997), Feature (1994), and White Columns (1992) in New York. He has participated in several group exhibitions including *Officina America* at Galleria d Art Moderna, Villa della Rosa in Bologna, Italy (2002), *Mood Painting* at Curt Marcus in New York (2000), and *Inside/Out* at Momenta Art in Brooklyn, New York (1996).

CRAIG KALPAKJIAN has had solo shows at Andrea Rosen Gallery (2002) and Robert Miller Gallery (2000, 1998) in New York, and Galerie Nelson in Paris (1999, 1997). He has been featured in numerous group exhibitions including *Bitstreams* at the Whitney Museum of American Art (2001), *010101: Art in Technological Times* at the San Francisco Museum of Modern Art (2001), and *Scanner* at the California College of Arts and Crafts, Oakland (2000).

PATRICK MEAGHER graduated from Harvard University in 1999 with an MA in Landscape Architecture and Urbanism. In addition, he studied at the Kunstakademie in Dusseldorf and received his BFA from Carnegie Mellon University in Pittsburgh. He had a solo exhibition at Riva Gallery in New York (2002) and has been in group shows at Artists Space (2000) and Lehmann Maupin Gallery (1999) in New York and the Lionheart Gallery in Boston (1998). Meagher has also curated a number of exhibits and is the founder of Artists Network Enterprises (ANE). He was a resident artist in the Lower Manhattan Cultural Council's Worldviews program in 2000.

JULIE MEHRETU received her BA from Kalamazoo College in 1992 and her MFA from Rhode Island School of Design in 1997. She has participated in group shows including *Casino 2001* at the Stedeiijk Museum Voor Actuele Kunst (2001), *Painting at the Edge of the World* at the Walker Art Center (2001), *Freestyle* at the Studio Museum in Harlem (2001), and *Greater New York* at P.S.1 Contemporary Art Center (2000). Most recently, individual projects have been presented at the Walker Art Center where she was an artist-in-residence (2002), White Cube in London (2002), The Project in New York (2001), and ArtPace in San Antonio (2001). She was a 2001 artist-in-residence at the Studio Museum in Harlem.

MATTHEW NORTHRIDGE received his MFA from The School of the Art Institute of Chicago in 1999 and studied at the Skowhegan School of Painting and Sculpture in the summer of 2000. His work has been included in the group shows *Emerging New York Artists* at the Arno Maris Gallery at Westfield State College in Massachusetts (2002), *Constructions* at the Arlington Center for the Arts in Arlington, Massachusetts (2001), *Just Like Heaven* at the Green Room in Brooklyn (2001), and *City Rythms* at the Pelham Center in Pelham, New York (2001).

SVEN PÅHLSSON has studied fine art in Stockholm and Oslo and recently received the Rune Brynestad Memorial Grant in Norway. He has had solo exhibitions at Spencer Brownstone Gallery in New York (2001, 2002) and the Oslo Museum of Contemporary Art (2001), and has been included in several group exhibitions including *Modellmakerne*, Riksustillinger, Norway (2002), *Animations* at P.S.1 Contemporary Art Center (2001), the Melbourne Biennale in Australia (1999), the Nordic Pavillion of the Venice Biennale (1997), and *The Name of the Place* at Casey Kaplan, New York (1997). Påhlsson has participated in a number of international video art festivals and was in the International Studio Program in New York.

ADAM ROSS received his BFA and MFA from the University of California in Santa Barbara. He has exhibited at Sara Meltzer Gallery in New York (2001), Nylon in London (2001), Shoshana Wayne Gallery in Santa Monica (1996, 1998, 2000), and Caren Golden Fine Art in New York (1997, 1999). He has participated in several group shows including *010101: Art in Technological Times* at the San Francisco Museum of Modern Art (2001), *By Hand* at Cal State Long Beach (2001), *Shifting Ground: Transformed Views of American Landscapes* at the Henry Art Gallery, University of Washington, Seattle (2000), and *Science Fictions* at White Columns in New York (1999).

DANNIELLE TEGEDER received her BFA from SUNY, Purchase in 1994 and her MFA from the School of the Art Institute of Chicago in 1997. Her solo exhibitions include *Love, Lust and Other Mechanical Systems* at De Chiara Gallery in New York (2002), *Undercurrents* at the Jan Cicero Gallery in Chicago (2001), and *New Work* at the Visual Arts Gallery at SUNY (1999). Group exhibitions include *Rumors of War* at Triple Candie in Harlem (2002) and *Wattage and Friendship* at mullerdechiara in Berlin (2001). She has received a Fulbright Scholar Grant (2002), a Yaddo Residency (2002), and the Marie Walsh Sharpe Studio Program award (2001–2002).

SHIRLEY TSE received her BA from the Fine Arts Chinese University in Honk Kong (1993) and her MFA from Art Center College of Design in Pasadena (1996). She has had solo exhibitions at Murray Guy Gallery in New York (2002, 2000), Shoshana Wayne Gallery in Santa Monica (2000), California College of Arts and Crafts in Oakland (2002), and Para/Site Art Space in Hong Kong (2000). She has participated in numerous group exhibitions including *Landscaping Ahead* at the Center for Curatorial Studies at Bard College (2002), *The World May Be Fantastic*, Biennale of Sydney, Australia (2002), *010101: Art in Technological Times* at the San Francisco Museum of Modern Art (2001), and *After the Gold Rush* at Thread Waxing Space in New York.

KEVIN ZUCKER received his BFA from Rhode Island School of Design in 2000 and his MFA from Columbia University in 2002. He has had solo exhibitions at Mary Boone Gallery (2002) and LFL Gallery (2001) in New York. Group exhibitions include *Open* at LFL Gallery and *Addiction* at The Arcade in Providence, Rhode Island (1998). He is the recipient of the 2000 Yvonne Force Prize and the 2000 Florence Lief Award.

HALUK AKAKÇE
Still Life, 2002
installation with two-channel video projection
dimensions variable
COURTESY OF BERNIER/ELIADES, ATHENS AND DEITCH PROJECTS, NEW YORK

RICCI ALBENDA
People Pattern, 2002
vinyl
dimensions variable
COURTESY OF ANDREW KREPS GALLERY, NEW YORK

AZIZ + CUCHER
Interior #5, 1998–99
c-print, edition of 3
50 x 72 inches
COURTESY OF HENRY URBACH ARCHITECTURE, NEW YORK

Interior #7, 1998–99
c-print, edition of 3
62 x 40 inches
COURTESY OF HENRY URBACH ARCHITECTURE, NEW YORK

Interior Study #2, 1998–99
c-print, edition of 7
19 1/2 x 14 inches
COURTESY OF HENRY URBACH ARCHITECTURE, NEW YORK

Interior Study #3, 1998–99
c-print, edition of 7
26 x 24 inches
COURTESY OF HENRY URBACH ARCHITECTURE, NEW YORK

Interior Study #5, 1998–99
c-print, edition of 7
20 x 20 inches
COURTESY OF HENRY URBACH ARCHITECTURE, NEW YORK

Glory, 2002
c-print
72 x 36 inches
COURTESY OF HENRY URBACH ARCHITECTURE, NEW YORK

NINA BOVASSO
Untitled, 1999
acrylic on paper
27 1/2 x 40 inches
COLLECTION OF DEREK ELLER & ABBY MESSITTE, BROOKLYN

Untitled, 2000
acrylic on paper
27 x 39 inches
COURTESY OF PAUL KASMIN GALLERY, NEW YORK

DreamHouse, 2001
acrylic on paper
30 x 40 inches
COLLECTION OF JOHN PANKAUSKI, WEST PALM BEACH, FLORIDA

T Ring, 2001
acrylic and ink on paper
30 x 40 inches
Private collection
COURTESY OF CLEMENTINE GALLERY, NEW YORK

VICTORIA HAVEN
Supermodelcity, 2000
tape, vellum, and pins
90 x 138 x 1 inches
COURTESY OF THE ARTIST

STEPHEN HENDEE
Silent Sector, 2002
site-specific installation of Cor-X, tape, and lighting
dimensions variable
COURTESY OF THE ARTIST AND HENRY URBACH ARCHITECTURE, NEW YORK

CANNON HUDSON
Serpent, 2002
oil on canvas
72 x 54 inches
COURTESY OF THE ARTIST

Colossus I, 2002
oil on canvas
72 x 54 inches
COURTESY OF THE ARTIST

Colossus II, 2002
oil on canvas
72 x 54 inches
COURTESY OF THE ARTIST

CRAIG KALPAKJIAN
Corridor, 1995
laserdisc
COURTESY OF ANDREA ROSEN GALLERY, NEW YORK

Room, 1996
cibachrome print on aluminum
29 1/2 x 39 1/2 inches
COURTESY OF ANDREA ROSEN GALLERY, NEW YORK

Stair, 2001
cibachrome print on aluminum
30 1/2 x 58 1/2 inches
COURTESY OF ANDREA ROSEN GALLERY, NEW YORK

PATRICK MEAGHER
UnitBead 2.0, 2000–02
user-navigable data projection installation
dimensions variable
COURTESY OF THE ARTIST

G4 Atrium, 2001
epson print
16 x 20 inches
COURTESY OF THE ARTIST

Tektronix Park, 2001
epson print
16 x 20 inches
COURTESY OF THE ARTIST

Wintel Splay, 2001
epson print
16 x 20 inches
COURTESY OF THE ARTIST

HP Way, 2001
epson print
16 x 20 inches
COURTESY OF THE ARTIST

JULIE MEHRETU
Retopistics: A Renegade Excavation, 2001
ink and acrylic on canvas
96 x 216 inches
DIMITRI DASKOLOPOULOS COLLECTION, ATHENS
COURTESY OF THE PROJECT, NEW YORK AND LOS ANGELES

MATTHEW NORTHRIDGE
New City, 1998–2002
paper and masonite, consisting of 3000+ individual
pieces
24 x 96 x 96 inches
COURTESY OF THE ARTIST

Untitled, 2002
collage
30 x 44 inches
COURTESY OF THE ARTIST

SVEN PÅHLSSON
with composer Erik Wollo
Sprawlville or Life at Highway Exit Ramp, 2001
digital video of 3D animation with sound
10 minutes
COURTESY OF SPENCER BROWNSTONE GALLERY, NEW YORK

ADAM ROSS
*Untitled (Too Far for the Eye to See, Always
at the Back of My Mind #2)*, 2001
oil, alkyd, and acrylic on canvas
48 x 60 inches
PRIVATE COLLECTION, NEW YORK

*Untitled (Too Far for the Eye to See, Always
at the Back of My Mind #3)*, 2001
oil, alkyd, and acrylic on canvas
48 x 60 inches
COURTESY OF THE ARTIST AND SARA MELTZER GALLERY, NEW YORK

Up on the Day, Behind the Sun #1, 2002
oil, alkyd, and acrylic on canvas
48 x 60 inches
COURTESY OF THE ARTIST AND SARA MELTZER GALLERY, NEW YORK

Up on the Day, Behind the Sun #2, 2002
oil, alkyd, and acrylic on canvas
48 x 60 inches
COURTESY OF THE ARTIST AND NYLON, LONDON

Untitled (Science Fiction 1), 2001
graphite on paper
23 x 29 inches
COURTESY OF THE ARTIST AND SARA MELTZER GALLERY, NEW YORK

Untitled (Science Fiction 2), 2001
graphite on paper
23 x 29 inches
COURTESY OF THE ARTIST AND SARA MELTZER GALLERY, NEW YORK

DANNIELLE TEGEDER
*Cream Under City with Secret Dome Forest and
Escape Transport Plan*, 2001–02
acrylic, enamel, and mixed media on panel
48 x 48 inches
COLLECTION OF EDITH AND JOSEPH DE CHIARA, NEW YORK

*Pink City Station with Dead-End Tunnels and
Lust Fire*, 2001
colored pencil, ink, acrylic, house paint, dye, pigment,
and enamel on canvas
48 x 48 inches
COLLECTION OF SUSAN EVANS, NEW YORK

*Garden Airline Escape Plan with Rhythm Transport
Center, Velvet Wells, White Circle Escape Plan Steamers,
and Hidden Signal Headquarter*, 2002
acrylic, flashe, and colored pencil on canvas on panel
48 x 96 inches
COURTESY OF THE ARTIST AND DE CHIARA GALLERY, NEW YORK

SHIRLEY TSE
Polymathicstyrene, 2000
extruded polystyrene
dimensions variable
COURTESY OF THE ARTIST, MURRAY GUY GALLERY, NEW YORK,
AND SHOSHANA WAYNE GALLERY, SANTA MONICA

Bionicpak, 2001
extruded polystyrene
2 parts: 15 x 48 x 33 inches & 31 x 48 x 33 inches
COURTESY OF THE ARTIST, MURRAY GUY GALLERY, NEW YORK,
AND SHOSHANA WAYNE GALLERY, SANTA MONICA

KEVIN ZUCKER
Untitled, 1999
acrylic, ink-jet and carbon transfers, and
enamel on canvas
68 x 56 inches
PRIVATE COLLECTION
COURTESY OF MARY BOONE GALLERY, NEW YORK

Angels; the Heads of Pins, 2000
acrylic, ink-jet and carbon transfers, and
enamel on canvas
60 x 45 inches
COLLECTION OF HUGH J. FREUND, NEW YORK
COURTESY OF MARY BOONE GALLERY, NEW YORK

World Without End, 2001
acrylic, ink-jet and carbon transfers, and
enamel on canvas
56 x 85 inches
COLLECTION OF DOUGLAS S. CRAMER, ROXBURY, CONNECTICUT
COURTESY OF MARY BOONE GALLERY, NEW YORK